BUILDING TECHNOLOGY 1

CONSTRUCTION TECHNOLOGY AND MANAGEMENT

A series published in association with the Chartered Institute of Building.

Each volume covers an important aspect of modern construction management. The series is of particular relevance to the needs of students taking the CIOB Member Examinations, Parts 1 and 2, but is also suitable for degree courses, other professional examinations, and practitioners in building, architecture, surveying and related fields.

Project Evaluation and Development
Alexander Rougvie

Practical Building Law
Margaret Wilkie with Richard Howells

Building Technology 2: Performance
Ian Chandler

Building Technology 3:
Design, Production and Maintenance
Ian Chandler

The Economics of the Construction Industry
Geoffrey Briscoe

Construction Management 1:
Organisation Systems
Richard Fellows, David Langford
and Robert Newcombe

Construction Management 2:
Management Systems
Richard Fellows, David Langford
and Robert Newcombe

BUILDING TECHNOLOGY 1

Site Organisation and Method

Ian Chandler

Mitchell · *London*

in association with the Chartered Institute of Building

First published 1988

Reprinted 1990, 1992

Typeset by Progress Filmsetting Ltd
and printed and bound in Great Britain by
Biddles Ltd, Guildford and King's Lynn
Published by The Mitchell Publishing Company Limited
4 Fitzhardinge Street, London W1H 0AH
A subsidiary of B.T. Batsford Limited

British Library Cataloguing in Publication Data

Chandler, Ian
Building technology.—(Construction
technology and management).
1: Site organisation and method
1. Building
I. Title II. Chartered Institute of
Building III. Series
690 TH145

ISBN 0-7134-5178-5

Contents

Part One
A CONCEPTUAL FRAMEWORK

1. Introduction

This is the first volume in a series of three dealing with building technology. It is concerned with the factórs that affect the methods and organisation of construction work, with particular reference to on-site activities. Volume Two, *Performance*, describes the factors and issues affecting the performance of buildings, such as: climate and its effect on buildings; building as a climate barrier; fire technology; building deterioration and specification. The final volume, *Design, Production and Maintenance*, draws together the three main aspects of the construction process and investigates their interrelations. A full analysis is also presented of the contextual issues affecting the choice of building technologies. A brief introduction to these issues follows this chapter.

How does one define building technology? A simple definition is not possible, and there are as many valid explanations as there are participants in the building process. However, there are a number of aspects which, when drawn together, give meaning to the term. These are shown in Fig. 1.1 (p. 15). The way in which these factors interrelate to create a matrix enables a building to be built or to be recycled.

Technology is dependant upon all the factors and they are not presented in any hierarchical order, since each must be seen as a contribution of equal importance.

CONSTRUCTION TECHNIQUES

The basic methods of creating and assembling the materials and components of construction form the core of understanding technology. The building technologist must understand the principles of construction and appreciate the most effective methods of realising the building element – foundations, walls, floors, roof and services.

What is the function of a damp-proof course? Where should it be placed? What materials can be used? These are the typical questions to which the answers should be known. How a dpc should be placed raises issues which are basic to the study of building technology. The assembly process will depend upon the skills available and the sequencing of activities. Temporary aids to production, such as access platforms, plant and machinery and site accommodation, will all affect the efficiency of an operation. The range of availability of resources open to the builder can dictate the construction technique. Safety issues can influence the way that an element is constructed, as in the need for earthwork support in excavations. The site layout can constrain the methods of construction and impose organisational difficulties. The builder will be responsible for organising site activities, and optimum methods of production will result from an understanding of basic construction details and from ensuring that the right resources are centred on an activity at the right time.

MATERIALS

There is a need to know the fundamental characteristics and performance of building materials and components in use. How does the manufacturing process affect the method of placement and ultimate performance? What role do new materials and components play in the development of building technology? To what extent does the introduction of a new material or component affect the methods of construction? When materials are placed adjacent to each other, are they compatible and do they function coherently? What factor most affects the choice of materials? Is it cost, availability, user requirements, function or performance?

SOCIETY

Every national and regional society creates its own needs and values. These will be reflected in the types and functions of its buildings. In turn this will dictate the types of materials and methods of construction. For example, western industrialised nations desire sophisticated, well-serviced buildings. There is a growing demand for leisure and sports facilities. The internal finishes and environment need to be robust to provide adequate performance in use. On the other hand, in developing countries the emphasis is on providing living accommodation with basic and minimal sanitary facilities. Flats built in Hong Kong have a concrete shell, with a small kitchen fitted with a work top and outlet for powering a cooker. The bathroom is small, with wash basin, toilet and shower. Living accommodation is one or two rooms, which may need to house up to seven people. This solution to housing the population is influenced by: lack of building space; the need to build many units to cope with an exploding population; the cultural requirement for a number of generations in a family to live together.

There are regional variations even in a relatively small country like the UK. Scottish society has developed its own range of values and legal framework which have influenced building technology.

STRUCTURE OF THE INDUSTRY

To understand the structure of the construction industry an insight is needed into the manner of building procurement. This insight is important for those practising in construction because it will enable them to act with integrity and ensure that buildings will meet their specifications. Those working in the industry also need to earn their living from it; to know how it operates is essential if adequate recompense is to be achieved. This applies both to individual salary/wage earners and to commercial organisations as a whole.

A study of the industry will show how a particular technological solution has developed and whether or not it can be improved by altering or involving others in the process. Would more objective scientific investigation improve the quality of materials? Who would fund this investigation? What would be the benefits to client and society?

The roles and responsibilities of people in the industry are currently changing and developing. If, for example, the architect's role changes to that of only producing the building's conceptual drawings, how will this influence the technology employed? Will the builder have to take greater responsibility for design and specification? Would this mean a greater willingness by builders to accept higher levels of responsibility? Who would generate building innovation – architects, builders, specialist contractors, consultants or material/component suppliers? How would changing roles affect the type and degree of education and training required for all functions in the industry? Site skills will be affected by technology becoming a process of assembly rather than a grafting onto basic materials of fixtures and fittings within the building.

LEGISLATION

Legislation can directly influence the technological solutions adopted. For example, the Building Regulation requirement is for walls to have a low thermal transmittance value. This value can be altered, which will affect the materials and methods of construction of external walls. The two half-brick skin cavity wall construction which was acceptable in the 1960s in the UK will now not meet the requirements.

Legislation spans structural stability, environmental health, planning, safety during construction, and noise on site.

Indirectly, legislation can affect a building's technology. For example, if a country imposed import controls, many materials and components might become unavailable, in which case alternative, homeproduced, items would need to be used.

Government policy can also influence the types of building produced. For example, the provision of grants for industrial manufacturing buildings would encourage their construction. Since many of these are built as single-storey lightweight steel-framed structures with lightweight cladding panels, an increase in this type of construction would occur.

CONTRACTS

In the UK the majority of construction undertaken by public authorities and private clients is let on standard forms of contract. Many of the clauses of these contracts can influence the technology used – for example, whether suppliers or specialist sub-contractors are nominated or not.

The contractual relationship between the client and the building team can also affect the choice of technological solution. For example, where the client chooses a package deal contractor he has to accept the building system developed by that company.

SERVICES

In recent years there have been major advances in the provision of services to buildings. Some buildings merely function as a shell for a mass of sophisticated service systems and in this situation it is the services which dictate the mode of building technology. Building technologists need an understanding of the basic service facilities and how they function, but overriding this is the need to know how they can be integrated into the process and the product of building. The role of services has become important in industrialised societies. Such societies could not function without service systems for power, water, internal transport, communications, sanitary hygiene, heating, lighting and ventilation.

DESIGN

The viewpoint of the building's designer can directly determine the ultimate technological solution. An architect concerned with saving energy will produce a building with high standards of insulation and limited use of service appliances. Designers with the needs of the building's users at the forefront of their considerations will seek to involve them in the design and build process. This may produce unusual solutions and make the construction process more complicated.

A designer who approaches a problem from a deductive viewpoint will produce a different solution to one who works inductively.

Techniques in design evaluation can be used to give greater credibility to the proposed building but their use is not widespread. When they are employed they can lead to a fundamental change in technology. Some of these techniques are directly related to the function of the building and to its elements, and attempt to analyse their value.

MAINTENANCE

Solutions to provide effective and cheap maintenance can be fed back for future design. The technology of construction, whether in detail, method of installation or substitution of materials, can be radically altered in the light of maintenance feedback. The methods of organisation of maintenance operations can help or hinder the feedback of information. Maintenance needs to be considered at the design stage so that the client is aware of the implications of the solutions offered.

A knowledge of how and why materials and buildings deteriorate is necessary during the design, production and use of a building so that appropriate solutions are formulated. Deteriorology is becoming a science in itself and hopefully will bring about a much better awareness of the way materials decay, fail or wear out.

Buildings are quite likely to undergo changes of use during their lives and to require substantial refurbishment. The methods of organisation and skills required for this can differ greatly from those in new building projects. The technological solution for buildings having the same function can be very different depending on whether it is new build or refurbishment.

PERFORMANCE

A building is usually judged on its performance. It is the ultimate aim of the design and construction process to produce a building which functions well with the minimum of maintenance. Standards need to be set to determine the criteria for performance. These standards should be based on the expectations of the building in relation to the environment in which it is located; the ambient climate may itself be able to protect the occupants from such hazards as fire. The materials, components, elements, services, construction detail, methods and amount of care in construction all contribute to the overall performance of a building. A failure in any one of these factors will not necessarily mean that the building as a whole is a failure, but one failure may lead to another if not rectified.

Buildings can be wasteful of energy and as this is relatively scarce and costly the emphasis must be on limiting its use, but still providing the required internal environment and services.

ECONOMICS

Dominating all technological matters and eventual solutions is money. Whether this is at a macro level, with national policy influencing design decisions, or at the micro level where the builder needs to make a profit, it is money which will determine many outcomes. One of the first questions a client nearly always asks is 'what will it cost'?

In capitalist societies buildings are an integral part of the monetary system, being regarded as investments or as security for loans. Their worth can be over and above that of the function they provide. During the construction process the values which the participants place on their contributions will differ, and in some cases might conflict, with a detrimental effect on the technology. These values need to be understood.

Even after considering all the above factors, it is still not easy to define building technology succinctly. Essentially, it is a study with many facets, covering all the factors which need to be considered in reaching sound technological decisions. The relative importance of each will vary from project to project, and can change with time during the course of a project. For example, it may happen that one material chosen will not be available or suitable because recent failures have demonstrated its poor performance, or that user requirements have changed, necessitating an alteration in technology.

The issues considered in the three volumes on building technology in this series do not pretend to be prescriptive or exhaustive, nor are they to be taken as immutable. There are some ideas and concepts which are based on fundamental requirements, such as performance, but their interpretation and development will alter in the course of research and practice. There is continuous publication of authoritative research papers, articles and technical literature worldwide which furthers the study of building technology. It is likely that further concepts not covered here will be forthcoming in the future. The three volumes provide an introduction to the relevant issues, based on current knowledge. They must be seen as drawing together the threads of building technology and not as giving specific answers. This is why questions are posed throughout the text. These questions have to be applied to real-life projects to generate viable answers. If there was only one correct answer to each question the built environment would not comprise so many different types of buildings and structures.

In this volume there is a brief introduction to the conceptual framework, encapsulating the process of determining technological solutions (a fuller discussion of these factors is presented in Volume Three). Before embarking on the discussion of building technology some reference is made to technology in general, and the role of building technology in particular. The way in which technological solutions are formed is governed by the constraints of site conditions and the aids available to carry out tasks. Building services play an ever-increasing part in the construction and recycling of buildings and their role and the problems of integration are discussed at length.

At the end of each chapter are a few questions. These are designed to extend and relate the contents of the text to experience and practice; it is unlikely that complete answers can be generated from the book alone. A vast range of literature is currently available, and new publications dealing with building technology appear weekly upon which an answer can be based. There is not likely to be one simple answer to each question. Indeed, one person may develop ideas and reach conclusions which are radically different from those of another, and neither will be wrong.

Finally, a number of building projects are presented and discussed in relation to the topics and to relevant contextual framework factors, to show how particular technological solutions were reached in specific cases.

2. Brief introduction to the contextual framework factors

In discussing the meaning of building technology a number of indirect factors were mentioned. These are shown in Fig. 1.1, interspersed with factors which are within the direct control and responsibility of the building team. For the purposes of clarity and definition the contextual factors are identified here. Only a brief introduction will be given as an in-depth discussion is presented in *Building Technology 3*.

1. FUNCTIONAL REQUIREMENTS

Every building project is primarily motivated by the need to serve a function. Houses are for living in; factories for producing goods; leisure centres for sporting activities. Each building is designed and built to satisfy particular client demands which can make it different from others of its type. Stemming from these demands are the standards of function and efficiency expected. Some of these standards will be determined by factors outside the control of the client and builder, such as environmental health requirements. Other functional standards, such as degree of service installation and quality of fixtures and fittings, can be set by the client.

In many cases the functional standards can be objectively measured. Two examples are the output of a heating system and the sound reduction capability of a floor. The heating system can be measured in two main ways: the amount of energy it uses and the heat/energy it can transmit for warming. The floor can have a defined decibel reduction factor. In certain situations these standards will be set by statutory legislation. A floor separating two users in a domestic building will need to be designed so that the minimal decibel reduction is achieved, but a floor (of possibly the same construction) in another building type will not be required to reduce sound transmission.

Standards of performance covering the major elements, components and services of buildings are common. As knowledge of materials, together with the way in which they are used, is increased, higher levels of predicting ultimate performance can be obtained. British Standards, Codes of Practice and Agrément Certificates have been given greater value in the 1985 Building

construction techniques

materials

society

structure of the industry

legislation

contracts

services

design

maintenance

performance

economics

safety

1.1 Indirect and direct factors in building technology

Regulations. If an Agrément Certificate has been obtained by a component, material or construction method then, insofar as it meets the Building Regulation requirements, it is evidence that the choice is suitable.

Standards of function are constantly changing; what was acceptable a decade ago may now be seen as inadequate. This change is not wholly a result of builders/architects producing new designs but arises from material and component producers introducing new products to meet society's demand. This demand can influence builders/architects in providing specific types of buildings. For example, a population of young single people requiring accommodation will create a demand for small, one-bedroom flats/houses. A population tending to have large families living together will create a demand for multi-bedroomed accommodation. In these cases the performance standards may not be objectively measurable, but if the accommodation is inadequate for society's need then, subjectively, as far as general standards of living are concerned, the users can justifiably criticise the performance of the building.

Function is not just the ability of a material or service to meet performance standards attributable to a discrete part of the building; it is also governed by the perceptions and needs of the society which generates the buildings.

2. SAFETY ASPECTS

The question of safety can be addressed to two areas: the process of construction, and the building in use.

Safety of site personnel during construction is of paramount importance and this goes beyond the wearing of hard hats. Falsework must be designed adequately. Safe access to and from the place of work is necessary. Health and welfare facilities must be provided.

The need to carry out work operations safely can have a profound effect on methods of production and in consequence building technology. For example, the need to prevent trench collapse could involve the use of linings and struts, but how will this affect the operation that is to be carried out in the trench? Might it not be better to use a method which does not require people to go into the trench? Can some other way be found to consolidate the ground?

The safety considerations of the building in use need to embrace two groups of people, those inside and those outside. There is a moral duty to provide safe structures, but how far is it necessary to go? Take, for example, the question of safety during fire. It is necessary to create a structure which will give adequate time, and the means, for people to escape. But should all buildings have some form of automatic fire-fighting equipment? What are the economics of this? Would money be saved from a reduction in the fire-fighting services? Who would pay for this equipment? Can life be equated with monetary considerations? The degree of safety and its implementation will influence the building technology.

3. ECONOMIC, PLANNING AND LEGISLATIVE FACTORS

A question which underlies the great majority of technological choices is 'What is the cost?' Put this way it does not fully describe the economic factors that will influence the process of construction, as cost itself is relative to the macro-economic milieu. This macro-economic climate is sustained and modified by government. Government fiscal policy can create an economic environment which may favour certain types of buildings, and may increase or curtail construction industry activity by controlling public expenditure or by changing the minimum lending rate. Grants for particular construction activities, such as home insulation, can boost work in that area of the construction market.

Micro-economic considerations pervade most human activities, whether it is earning enough money to live on or, if in business, to generate capital and profit to ensure the viability of the enterprise. Therefore, whether it is the client spending money or the builder earning it, there is a need to determine and control costs. In the final analysis it is quite likely that the choice of construction method, materials, components, standard of finishes, etc. will be determined by cost.

The creation of the built environment is increasingly influenced by planning legislation. As societies become more interactive, and towns and cities change and grow, some order is needed to protect all the people living and working there. We may debate how far this control should be extended, but there are obvious cases where certain building types are totally unacceptable in specific locations. The debate on aesthetics rages with great force, as styles of architecture are a matter of personal taste. Planning not only embraces type, position and aesthetics of buildings but, in the context of building technology, such aspects as public health and government policy.

Legislation can directly affect the process and product of building and there are many Acts on the statute books which control building. In addition a good deal of other legislation, drafted on topics not directly seen to be of interest to construction, has over time influenced its technology. Much economic legislation comes within this category. From time to time this legislation will be altered, repealed and added to: the building technologist must ensure that he is familiar with the current requirements.

4. RESOURCE AVAILABILITY

Ensuring that the appropriate materials and components specified are available is a constant problem. Overriding this consideration is whether the specification is drawn up in full recognition of its effect on other factors. Where does the material come from? If imported, what might its effect be on the nation's balance of payments? Can it be effectively maintained? What is the overall cost of the item until the end of its useful life?

The prime mover of all construction activity (together with most other manufacturing and business activity) is energy. Since the oil crisis of the early 1970s there has been a greater public awareness of the finite nature of fossil fuels. Alternatives accepted by everyone as safe have not yet been satisfactorily produced, and whatever energy source is used in the future will affect the process of building as well as the materials and components. Building design may have to be adapted to suit a new energy source such as solar energy.

5. APPROACH TO DESIGN

The divisions between client, designer, builder and user can have a detrimental effect on technology. The distance in time and levels of communication between client and user (if not one and the same) can adversely effect the effective performance of a building. Different approaches to building procurement endeavour to mitigate these problems by creating closer harmony between all the parties to the construction process. Who is responsible for the building's performance? How can this performance be monitored? What are the lessons for the future?

The way in which an architect designs a building can influence its technology. Is it designed from a functional viewpoint or to follow an aesthetic style? Is the concern with the whole building, or is the building the sum of its parts?

6. TECHNOLOGICAL CHANGE AND DEVELOPMENT

The last few decades have seen phenomenally rapid change in technologies. The building industry tends to use developments from other sciences and technologies to enhance its process and its product. Many new materials and components come on to the market and the choice is ever-widening. Advances in power tools, plant and equipment alter work practices. The use of computers will undoutedly affect the process of construction – will some buildings be constructed by computer-controlled robots? How will this alter the respective roles of builder and architect?

Technological development is the result of research. Who does this research? Should research be increased, or redirected? To what extent is the dissemination of research results effective? How far should research and development be linked with education and training?

By being aware of current trends and reviewing past developments it is possible to gain some idea of possible changes in building practice and products. It is not possible to predict the future, but there are discernible trends which, if analysed, evaluated and synthesised, will be of benefit to the builder, and ultimately to the client. For example, the introduction of small tractor-type diggers and front bucket machines has enabled productivity to be increased in large-scale refurbishment projects. These machines are light and manoeuvrable and can be employed on strong upper floors of buildings, or for the movement of materials, so enabling premixed mortar in bins to be transferred directly to the point of use. This cuts out the need for on-site mixers and for labour to operate them and carry the mortar to the bricklayer.

7. SOCIETY/TECHNOLOGY INTERFACE

The architect is at the meeting point between society (client) and technology (the building process). Generally, it has been the architect who interprets the client's wishes into a design, and who has to utilise the best technological solutions to meet the client's objectives. Increasingly, however, the builder is carrying out this function of interpreting society's needs. Indeed, many private house speculative developments are not architect-led at all, but are a direct result of the builder relating to the needs of society. Another instance of the builder undertaking this task is in the client's use of project management teams. Many of these teams are commercial organisations spawned from builder/contractors. Some are led by individual professionals who co-ordinate the design team and builder and sub-contractors. In each case the architect is not the prime contact between society and the industry.

There is much public debate as to whether the construction industry meets needs in a satisfactory manner. Unfortunately, there are many disreputable builders, and even respected designers and builders do not escape censure, which is justifiable in some instances. Society has two legitimate criticisms. One is that the quality of construction, its materials and standards are inadequate, and the other that builders take too long over contracts. They arise from society's contact with small builders, many of whom have not been adequately trained or educated. In 1987 the UK government initiated a major investigation into 'cowboy' builders, their service to the public, the possibility of their evasion of taxes and their contribution to the training of new entrants to the industry. It remains to be seen if this inquiry will have an impact on the quality of construction.

Another initiative, taken in 1987, was in conjunction with the United Nations International Year of Shelter. A campaign highlighted the declining standard of dwellings in the UK (and the world) and brought pressure on government, local authorities and the industry to ensure that housing facilities and standards are meeting requirements. Launching the campaign, Lord Scarman stated that there were one million dwellings technically unfit for human habitation. Are these a result of people not caring about buildings, preferring to spend their money on other things? In other words is this state of disrepair brought about by society's neglect rather than by the construction industry? If society spent more money on better materials and methods of construction, might not standards improve dramatically?

Perhaps it is the relationship between society and the building industry which has the profoundest effect on building technology, which in turn produces the built environment. One measure of the values of a society is its built environment.

8. USER VALUES

Many users of buildings have no contact with client, designer or builder during the construction period. Some users may become involved in the process at an early stage, or later, but even then can exert little or no influence over the technology employed. In the UK most private house development is based on standard types, the only variation being in colour schemes, or perhaps in kitchen and bathroom fittings. The customer usually buys from a set plan and during design the architect does not cater for personal idiosyncracies. Much commercial building is undertaken by clients who use the activity to make money, either by letting the property or selling it outright. Again, at design stage there is no research into the ultimate user's requirements, but the designer has somehow to create an environment which is expected to meet user values.

The members of the building team approach the construction project from different perspectives. For example, the client may see the building as a financial investment and source of revenue; the architect as a project to enhance his reputation; the builder as a means of making a healthy profit. Ideally, the three views should not be in conflict, but there are instances where ill-defined objectives or extreme divergences cause conflict. In this case the technology will suffer and the performance of the building will be below that expected.

The final verdict on the real value of a building can only be given by the people that use it. The Pompidou Centre in Paris, France provides an example of measurable user satisfaction: this building now receives more visitors than the Eiffel Tower. This could be attributed to the design meeting the user's requirements as regards function and aesthetics. It has an interesting architectural style which attracts people, but also fulfils the main function of allowing entertainment to flourish. The technology is appropriate.

SUMMARY

A fuller description of the foregoing eight contextual framework factors is given in *Building Technology 3*. When reading the subsequent chapters of this volume these factors should be taken into consideration. Although direct reference to them might not be made, their influence is implied. Technological decisions are not taken in a vacuum; they emanate from society and are bounded by constraints existing in that society, or impinging upon it from the wider world. Many of these influences may not be consciously felt, but exist within a person or group as they are and have developed within a culture. For example, in the UK there is strong social desire to own a house. The common type of house is a two-storey, three-bedroom dwelling, with lounge/dining room, kitchen, bathroom, central heating and garage, situated on an estate away from a town centre, but close to amenities such as schools and shops for basic foods and commodities. 'An Englishman's home is his castle': there seems to be a cultural 'need' to have four solid walls enclosing the living areas, therefore, and traditional materials such as bricks (lately various types of building blocks) are thought to meet that requirement. In the late 1970s and early 1980s a number of building developers started selling houses constructed with a structural timber frame. In many cases brick was used as a cladding, not only to give a pleasing appearance, but also to signify solidity. The commercial advantage of this form of construction was its speed of assembly: the average brick-built, 'wet' internal finish house takes about four to five months to build, whereas with a timber-framed house construction time can be reduced to three months. This meant that building developers could achieve a faster turnover of dwellings, with a consequence quicker return on their investment. It is possible to install the foundations up to and including the ground floor slab prior to a purchaser contracting to buy. When a buyer does make a commitment the house can be built and ready for handover six to eight weeks later. As the majority of materials, components and frames are bought from suppliers, by the time they render their invoices the developer will have received the payment for the house and will be able to settle the invoices immediately. Generally, this attracts a small percentage discount for prompt payment. The developer will only need to finance the whole of the work up to ground floor slab (plus, of course, general roads and services work) and the labour costs of completing the house. Compared to a traditional brick load-bearing structure the capital outlay can be much less; if the money is borrowed it will cost less. This saving can generate extra profit, or can be used for further investment in, say, land for building, or it can be passed on to the consumer in the form of improved standards, extra amenities in the house or a lower purchase price.

As a result of an investigative television programme which criticised timber-frame dwellings as being susceptible to decay and rot if not properly treated and constructed, the number of houses built using this system fell dramatically, from a rising 28% of the total to its present 5% (1987). A major housebuilder's share price fell considerably, causing a reduction in its development programme owing to lack of capital. The technology of timber frame was discredited, a factor which the manufacturers of bricks and blocks were quick to exploit. An extensive advertising campaign extolled the virtues of traditional bricks, together with blocks giving a sound, solid construction. Owing to the development of higher thermal resistivity in building blocks the cavity wall could also meet the 'u' values reached by timber frame (in combination with glass fibre insulation and vapour barriers). Thus a major selling point for timber frame, its high thermal insulation giving a comfortable and low cost energy environment, could now be matched by a brick/block cavity wall.

Analysis of the rise and fall of timber-frame housing construction can illuminate a number of contextual framework factors. It demonstrates the interrelationship between contextual framework factors and building technology. Society's values set the demand for houses, which are then built to a layout and style appealing to the needs and aspirations of the purchasers. Their construction is governed by the economics of the developers, who in turn are influenced by the national economic climate. For example, a large increase in minimum lending rate will increase the mortgage rate, which in turn will reduce the demand for mortgages, which then reduces the number of new houses demanded, which then affects the level of business activity of the house developer. Planning and legislation directly govern all house building and its technology. Functional requirements, such as space heating, will need to be catered for in the design of new houses. The demand for more sophisticated services will influence the approach to design and the role of designers. Research and development has generated new building blocks which have the strength and high levels of thermal insulation to meet the requirements of users and statutory legislation. The issue of safety also led to a decline in the number of timber-frame houses – was the timber strong enough? could it adequately resist rot and beetle attack? – because if it did not this would lead to the collapse of the house. With a sharp switch from timber to brick a greater demand was placed on brick manufacturers. Brick stocks were diminished and some types of bricks were not readily available, which led to delivery delays. The factors operating at the society/technology interface were brought into full play in this example.

QUESTIONS

1. Evaluate the influence of contextual framework factors on the use of lightweight sheet metal cladding on new factories.
2. Discuss whether or not the construction of buildings will be further constrained in the future by contextual framework aspects. Should it be?

3. The Role of Technology

One dictionary defines technology as 'a discourse or treatise on the art or arts; the scientific study of the practical or industrial arts' (*The Shorter Oxford English Dictionary*). This two-part definition shows that a change in meaning has occurred over the years, from its being centred on 'the arts' (such as literature, painting and music) to its being commonly accepted as applying to the 'practical' or 'industrial' arts. Examples of these arts are the making of porcelain and the production of motor cars. In the definition, the phrase 'scientific study' should not be forgotten. Analysing and evaluating technological activities is part of the process, since they do not become meaningful until they have been understood. Technologies are produced by groups of people: they are co-operative activities within societies. Study of a technology presupposes some knowledge of the characteristics of the individuals or groups of people who will practise or apply the technology. These people, in their interaction with others and with the technology, are shaped by the social, cultural, political and physical environment within which they live. Therefore, no technology can be separated from the environment from which it emanates or within which it might be assimilated.

The range of sophistication of technologies has grown with the overall development of mankind. Indeed, it can be said that mankind has developed because of technological improvements; but what is meant by 'development'? Can it be measured in the time taken to travel, which has certainly been reduced, or in the levels of comfort that can be found in living accommodation? Development can also be measured with respect to the level of civilisation and the value placed on human life. Health care has seen dramatic improvements, mostly due to advances in technology, but there has also been the rapid and no less dramatic development in the means of destroying mankind with nuclear weapons. Two world wars this century have not shown any improvement in man's ability to live peacefully or to control aggression. Has technology failed here?

A technology which can illustrate these issues is that of passenger aircraft. Fig. 1.2. gives a schematic demonstration of the development of passenger aircraft. In parallel with the improvements in the actual aircraft, its airframe and engines, are

	biplanes	petrol engines
grass strip		
sheds	monoplanes	wood frames
		canvas cladding
hangars	seaplanes	metal frames
concrete runways	bombers	metal fuselage
radar	fighters	
control tower		jet engines
terminal building	passenger planes	galleys
safety services	jumbo jets	automatic pilot
navigation beacons	shuttle	self landing system

1.2 The development of passenger aircraft

developments in airfield construction; control of aircraft on the ground and especially in the air; air crew training and passenger service; fare prices; passenger embarkation points; national identification with airlines; number of passengers carried in one aeroplane – all of these have encouraged a demand for air travel. A further possible development beyond Concorde would be an aircraft to travel on the boundaries of the earth's atmosphere, virtually a spaceship. Long-distance flight times could be halved, with economies in fuel as well, if this idea was realised. Safety, too, has always been uppermost as a factor in air travel and it takes on a greater import when large numbers of passengers are carried in one aircraft; a serious accident could entail heavy loss of life. Public interest is centred on safety aspects and many people will not travel by plane because of their fear of an accident. Whilst aircraft production and improvement is the central line of technological development this is not taking place without reference to the factors mentioned above. With respect to safety in particular, after the fire on board an aeroplane at Manchester Airport in 1985 the use of hoods for passenger protection in case of fire and its accompanying smoke and fumes may become commonplace on all aeroplanes. Although it is only a relatively small number of years since the first powered flight the technology has progressed rapidly, in sharp contrast to the development of building technology.

A BRIEF HISTORY OF BUILDING TECHNOLOGY

Shelter has always been one of man's prime needs. Archaeological evidence shows that caves and rock shelters were used for living in tens of thousands of years ago.

The first type of construction that employed a knowledge of materials and structure was that of wattle, wood, branches, grass and, in colder climates, turf. These dwellings used the materials that were close at hand and an expertise developed which was handed down by word of mouth and by demonstration. They were simple in shape and usually circular, still the most structurally strong shape. Wattle or branches were interwoven to produce the walls and roof, giving a supporting structure for the covering of leaves, grasses or turf. In areas where there were rocks or boulders a dwarf wall was built upon which the wood structure was placed. In some cases the floor of the dwelling was sunken. Few tools were required: structural timbers were chosen to meet exact lengths and the coverings could be cut if necessary by simple stone or metal tools.

A few thousand years ago civilisations developed tools and techniques for cutting and shaping stones: those societies were centred in the eastern Mediterranean. This technology coincided with the development of metals and of a large labour force (usually slaves under the control of skilled master architects and craftsmen). Evidence of this type of construction is now seen in ruined temples, palaces, shrines, tombs and cities. Houses and shops were built of stone in the larger settlements, demonstrating that the culture could support a high level of building technology. The degree of accuracy in cutting and dressing the stones obtained by some cultures cannot be surpassed today, even with modern powered tools. The method of jointing blocks and forming arches cannot be bettered. It is obvious that a great deal of effort and pride in workmanship went into the construction of these buildings.

In Europe the Greeks, followed by the Romans, created the art of architecture, with the Romans furthering the science of building. They developed mortars and concrete and clay roofing tiles. In these cultures reliance on slave labour to carry out the building work declined, to be followed by the emergence of wage earning craftsmen and labourers. Finishes to houses were applied, such as stucco, which was decorated. Mosaic tiles were used on the floors. Although there is little actual evidence of the types of lifting machinery it was obviously sophisticated and efficient to lift the heavy stones into place. The Romans developed the technique of vaulting, springing from columns, to give large clear spans. After the collapse of the Roman Empire, a general standstill and even degeneration occurred in building technology. The next leap forward did not occur until around 1100, with stone castles being constructed in Europe and the Middle East. This defensive building reached its height during the Crusades. The European Renaissance brought in its wake an interest in constructing buildings which were more than functional. Decoration, both inside and outside, was extensive and wealth was reflected in grandeur of scale. Training of new craftsmen was organised by guilds, which reflected the growing complexity of society and the value it gave to the built environment.

In the period 1750 to 1850 a greater understanding of the manufacture and use of cements, mortars and plasters was achieved. The organisation of the construction process, and the industry itself, began to change. The employment of trivial trades, the emergence of the building contractor as an employer, and the independence of the architect grew at the expense of the master mason. The client did not directly employ the tradesmen, but entered into a contract with a contractor who took over that responsibility. Sole traders, partnerships and limited liability companies organised and controlled the building work to the design of an architect. Many builders carried out speculative developments, these not being under the control of an architect or client. Both the Chartered Institute of Building and the Royal Institute of British Architects were founded in 1835. Craft guilds were still strong during this period, having started in the late 1500s. They provided most of the training for apprentices. Trade unions grew and exerted a strong influence over work practices and wages. A clear division between design and construction developed, with further subdivisions into specialist trades, craft guilds and unions. It was during this period that the Industrial Revolution occurred in the United Kingdom, which carried in its wake the need for a massive increase in houses and factories. The majority of these buildings were constructed in load-bearing brickwork, generally one brick thick, with stone flag ground floors, timber joist and board upper floors and a slate-covered roof supported on timber rafters. The face of internal walls was plastered. Sanitation was supplied by a water closet in the back yard of the house, water from mains under the road serviced a sink and the toilet, and sewers carried the effluent away. Heating and cooking was based on a coal fire in a cast iron range.

The standard of domestic comfort increased rapidly from the end of the 1700s to the middle of the 1800s.

From 1850 to 1900 the development of the framed structure allowed larger buildings to be erected. Large-span roofs covered railway stations, multi-storey buildings rose in the United States of America. Initially, the frames were in cast iron, such as the Crystal Palace built for the Great Exhibition of 1851 in Hyde Park, London. As the century came to an end, steel technology was improved and this enabled strong structures to be built. The balloon frame in timber was developed. Stone as a load-bearing wall material was virtually superseded by brick, being used only for decorative external finishing. Gas lighting was commonly used in the average town house and as street lighting, quite quickly to be replaced by electricity. Many small and medium-sized buildings were prefabricated in timber or cast iron and shipped from the UK to its colonies around the world: factory production techniques were being adopted (for a detailed history of factory-produced buildings see Herbert, *Pioneers of Prefabrication*).

The use of concrete increased, with the addition of steel to give tensile strength. Further improvements were made in the manufacture of cement, with consequent effects on strength and durability.

Matching high-rise development in steel was the new technology of lifts. The lift enabled people to go to any height, so that buildings were not restricted by human physical limitations. The design of the building's plumbing and other services kept pace with the increase in height.

From 1900 there have been many advances in the technology of building, primarily in the use of materials and in the levels of internal comfort. On-site techniques do not appear to have kept pace, however. Speed of construction has not greatly improved, and there are still major problems in achieving good quality constructional detail.

Some important technological innovations are listed below.

Steel: wider range of steel sections and profiles, development of reinforcing bars and prestressing techniques.
Composite cladding panels.
Sanitary ware in plastic.

Concrete: improvement in understanding of concrete; slip form techniques of precasting; large panel construction.
Fire protection: to the structure, and in fire-fighting services and means of escape.
Services: more efficient means of heating and ventilation, faster lifts.
Structure: shell roofs; tension structures; diaphragm walls; space frames; piles.
Production: earth-moving plant; cranage; small power tools; geotechnic techniques.
Plastics: external claddings; components and fittings; surfaces; insulation.

The application of these innovations has been accompanied by research into properties and performance. Scientific investigation has provided many answers regarding the structure of materials, their compatibility and uses. Social science and management studies have shed light on the ways and means of integrating men and materials to create efficient technology. Education and training has developed from craft level, to degrees in Building and postgraduate research. Organisations such as the Chartered Institute of Building are keeping pace with change by publishing learned papers and providing courses of study suitable for the demands of the technology. The construction industry comprises a multitude of interests: there are employers' federations, trade unions, professional institutions, many disparate job functions and divisions. Whilst scientific research in universities and commercial laboratories tends to lead the way in the development of materials and techniques, the discipline of building technology in its application to production is growing in polytechnics. There is as much concern with the study of the industry with respect to its effective functioning as there is with the development of new techniques and materials. Change is continuous and rapid, so much so that many professional institutes concerned with the built environment positively encourage their corporate members to keep up to date by undertaking continuing professional development programmes.

The latter half of the twentieth century has brought rapid change to the construction industry. New materials, methods of construction and ways of organising the construction process have been introduced and are still being developed. It is unlikely that the speed of change will falter for some years yet.

RELATIONSHIP BETWEEN TECHNOLOGY AND SOCIETY

Technology can be of great benefit to man. The majority of people would say that technological advances have made life easier and more comfortable. Others, rightly, question some of these advances and point out that they can have detrimental effects. Take the case of motor cars. They enable people to travel quickly, comfortably and in their own time, but in so doing, roads run through the countryside, the atmosphere becomes polluted and accidents exact a heavy toll in injuries and loss of life. Whilst manufacturers strive to improve standards of comfort, performance and safety in cars they do not always take the initiative in doing so. For example, some nations have passed legislation to reduce the emission of exhaust pollutants. Prior to this legislation car manufacturers had not bothered to introduce higher standards of emission control. This begs the question, would they ever have done so without government intervention?

A fuel, namely oil, needs to be extracted from the ground, processed and distributed. Those nations possessing oil reserves can use the money from its sale to invest in other technologies. Those without reserves have to import the oil and this will affect their economic policies and standards of living. High oil prices could lead to a lower standard of living for them, whilst lower oil prices could bring about a lower standard of living for those nations dependent on selling oil. A major benefit from oil-fuelled engines can be found in agriculture. Marginal land can be ploughed and food harvested efficiently, which lowers the cost of food and allows production of enough for all mankind. According to some commentators, there is enough food in the world to feed everybody adequately – starvation should not exist. Unfortunately, these food sources are not always available where they are most needed. Economic and political factors determine, to a great extent, the distribution of food; technology can provide the solution but its implementation can be restricted.

There are advantages and disadvantages to the use of oil as a fuel. Today, it is a commodity which forms the basis of many national economies, whether as producers or buyers. In the future, as the oil reserves run out, another commodity could underwrite the economic viability of countries. Might it be access to nuclear power? Might it be something that comes from the sea, such as natural gas in the sea bed silts?

Man has developed by the use of science and technology. To differentiate between the two can be a complex task, but generally science is concerned with the investigation of matter and phenomena, and technology with applying science to create useful objects and processes. In a study of the history of building technology it is noticeable that most advances have arisen on the back of scientific knowledge. Improvements in cements and mortar were a result of research in laboratories. Of course, some advances came from empirical, in situ invention, such as large-span timber roofs, but most technological improvement in building is not as a direct result of architect's design or builder's work on site but stems from quite remote industries or scientific investigation. Glass fibre, for example, has been developed elsewhere and adopted for some uses in construction.

Bowley, M. (*The British Building Industry: four case studies*, 1963) has shown that the promotion of new and varied construction materials and techniques was by people and organisations who were interested in producing goods and services for building. They could visualise a use and a market for their products and, therefore, invested in ways and means of producing and selling their innovations. Generally, designers and builders did not themselves develop the initial ideas; building systems used in the UK in the 1960s for dwellings were bought in by builders from firms not directly in the contracting field. But it is the designers and builders who accept the credit for successful innovations and sometimes try to minimise their responsibility when things go wrong. Society sees the builders and architects as being responsible for the technology, even though it might have been adopted as a complete package from a third party. This assumption is not unreasonable as the designers and builders govern the process of construction. The designer's specification determines the choice of materials and components and the builder's work the levels of quality and performance. They both act on behalf of the client, and the client looks to them for advice and the production of the building, irrespective of the system and whoever is employed under it. This singular relationship is changing, however, as the technology of building changes.

Before this point is illustrated some links between technology and society (their relationship and respective roles) should be put into historical perspective.

The development of building technology, already briefly outlined, will now be placed in the context of the interface between society and technology. Fig. 1.3 overleaf shows the structural development of buildings, which is closely allied to the change from insecure to confident societies.

An important function of buildings for the ancient Egyptians was their role in immortalising the ruling dynasty and in the observance of religious beliefs and rites. Structures were symbols of power, in both the physical (subjugation) and spiritual senses. There was little concern for the buildings of the common population – or for life during the construction process, the majority of works being carried out by slave labour, easily

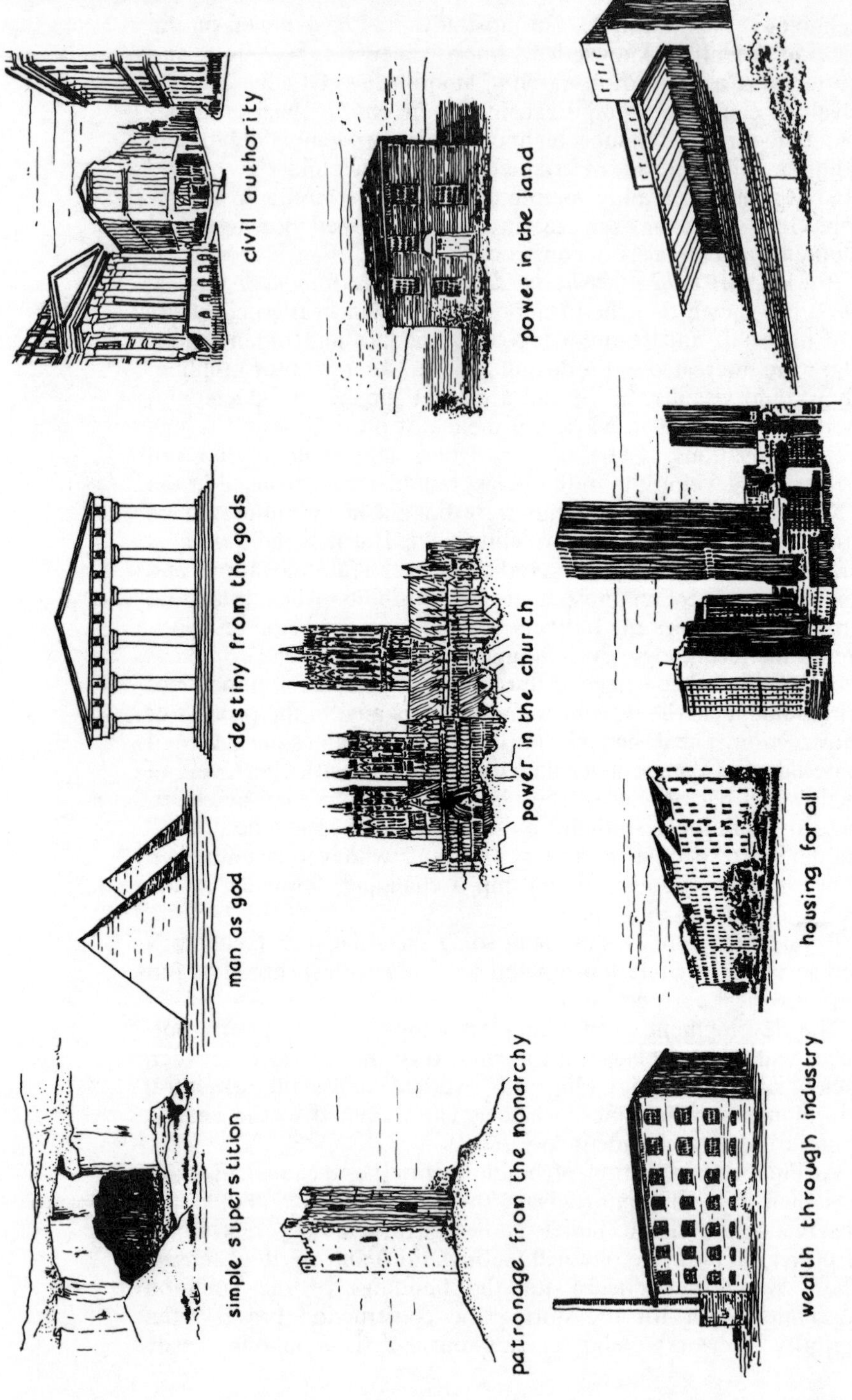

1.3 *The structural development of buildings*

replaceable in the eyes of the Egyptians. There is evidence that Egyptian towns were built in stone, but the buildings were small in area, abutted to each other and were not plastered or decorated. In comparison to the grandeur and decoration of the tombs and temples and palaces the town dwellings were insignificant and purely functional. The role of building technology was to impress the population of Egypt (and the rest of the Mediterranean peoples) and to reflect the power of the rulers. Its essence was primarily spiritual.

The Roman Empire saw a move away from purely spiritual reasons for building. Temples and palaces were still built but much time and effort was put into the construction of facilities for war and defence. There was also a need to service large town populations and rich families. Consequently, there were improvements in hygiene facilities and town layouts. Fresh water was ducted from afar to distribution fountains and even to individual houses and baths. Streets were paved and gutters formed to direct effluent. A major innovation of the Romans was central heating: hypocausts provided a system of underfloor and wall heating. Open fires were undesirable as the control of smoke and the distribution of heat was a problem, so instead stone floors were supported on pillars of tiles and the hot air circulated between the floor and the ground; hollow tiles took the heat up the walls. As the floors were not in direct contact with the ground the problem of rising damp was mitigated. Even so most dwellings at this time still only had a beaten earth floor.

Roman society respected designers and constructors of buildings and some of them prospered. Craftsmen were allowed to ply their trade freely and compete with each other. Stonemasonry, carpentry, plastering/stucco and mosaic-laying were the common trades, with obviously some skills in plumbing as shown by the number of lead and timber pipes found. The Romans used builders to help them invade and subjugate other peoples. Walls, forts, barracks and centres for administration were constructed initially in timber but quickly replaced by stronger and more durable stone. Houses of rich farmers or dignitaries showed a high standard of materials and finishes: internal walls were plastered and decorated with paintings, floors covered in mosaics, while fountains and ponds were a feature of internal courtyards.

After the fall of the Roman Empire, Byzantium flowered briefly. Art and architecture were applied to buildings to give symmetrical form and intense decoration, but little or no advance in technology was discernible.

The re-emergence of building as a technology did not occur until around 1000 AD, following the spread of Christianity through Western Europe. It was characterised in its later stages by a distinct architectural style based on Islamic cultures in Eastern

Europe and the Middle East. Stone was the predominant material and internal decoration was sparse. Churches, cathedrals and castles were built, reflecting the interests of the state. It was during this time that carpentry moved from being primarily concerned with the structure, walls, floors and roofs to the provision of shaped, moulded and panelled doors, screens and facings to internal walls. The joiner became a craftsman in his own right. The average dwelling was still a simple affair. Open fires provided warmth, and chimneys and flues controlled ventilation and smoke. As feudal society developed, the role of the castle moved from a strictly defensive one to that of impressing both the local population and other lords. Churches and cathedrals were built to the glory of God and to show the growing power and influence of the Church.

Until the nineteenth century, building served the ruling elites. As the ordinary person slowly gained influence, and peace rather than war became the norm, the technology of construction was directed more towards the dwelling house. Bricks were easily and cheaply made, enabling large houses and palaces to be built. In towns the timber frame still predominated but walls were infilled with brick panels. Glass was improved and enabled windows to be placed where required, but only in small sizes. As towns grew, so did the craft guilds. The guilds controlled entry to the trades and also rates for workers.

Training was carried out under the supervision of an experienced person and was entirely 'on the job'. Technology was beginning to benefit an increasing proportion of the population. Standards of comfort were improving with the use of brick and stone or clay roof coverings. The elements could be kept at bay satisfactorily. The building crafts as a social activity were highly respected by the population, and master craftsmen held the highest status below landowners and members of the governing class. The building crafts were the leading technology until the Industrial Revolution swept through the British Isles, starting in the late 1700s.

Building technology during the period of the Industrial Revolution was a means to an end. It provided the basic structures for the manufacturing industries and for housing the population explosion and the drift to the towns. Cast iron and wrought iron enabled large spans and great heights to be reached. Brick, timber and slate were the basic materials for artisans' houses: the load-bearing brick wall was the most common form of construction. The increase in the number of houses in towns demanded that attention be given to standards of hygiene. Toilets, of the night soil type, were provided in back yards. Piped water was laid to many houses. Chimneys and flues allowed ranges to be installed for cooking and heating. Sewers in the roads were built towards the end of the century, allowing flush toilets to be used.

Building in the United States of America reflected the country's dynamic economic growth and pioneer spirit. Using steel frames multi-storey structures became possible. Shops, offices and factories were built to demonstrate wealth and the state of the art in building technology. Building not only provided functional space for commercial activities (so that a profit could be made from those activities) but were also a symbol of that enterprise. The role of building technology was to exalt other technologies and sciences and to provide the basic facilities for their advancement. Building technology now followed where others led.

In conjunction with the swing in emphasis of building technology, from being an end in itself to a means to other ends, was a change in the organisational relationship of its participants. The building contractor emerged as an autonomous entrepreneur employing labour, purchasing materials and directing the works, to the design of an architect. The architect acted directly on behalf of the client. The builder as developer became commonplace, primarily buying land and constructing houses for rent or sale. The strength of the craft guilds diminished, except in upholding the levels of skills and training procedures. The rights and benefits of the workers were, intermittently, looked after by the trades unions. Employers joined together in federations. Architects were divorced from the day-to-day activities on site, their role becoming predominantly one of design and specification via drawings and written documents. The architect became a respected professional member of society, seen to be above mere commercial considerations. The status of the craftsman and the building contractor diminished as other industrial artisans and entrepreneurs gained respect. A few builders, certainly, were guilty of shoddy building and of exploiting the market for houses, demanding large rents for inferior dwellings. These instances provided material for newspapers and therefore became national knowledge: consequently the reputation of builders was permanently sullied.

In the UK, multi-storey factories up to eight storeys high were constructed, but they did not reach the heights of those in the USA. Steel technology in the USA outstripped developments in the UK. Why did not UK buildings reach the heights of those in America? Firstly, the client and general society did not like nor trust tall buildings (or their builders?). There was and still is a preference in British society for buildings which do not dominate the environment. Secondly, land in towns was relatively cheap and extensive so buildings spread out rather than up. Thirdly, the technology of building high was not as developed in the UK. With little or no advanced steel production in structural members, nor a significant body of knowledge about the characteristics of steel and concretes, it was not possible to build high with confidence.

Fourthly, there was a large pool of people skilled in the traditional crafts such as carpentry and bricklaying, with little or no provision for the training of new skills.

The technical institute which offered evening courses and lectures concentrated on the new sciences and technologies for the manufacture of goods. A building apprentice, on the other hand, learnt at the elbow of a craftsman who could only pass on what he knew, so there was no opportunity to learn new skills or methods. It was not until an apprentice attended a college that he had the chance to explore the advances in techniques. This was not common until the middle of the twentieth century.

At the beginning of this century the roles of builder as contractor and architect as designer had become firmly established. There had been little development in the technology of building. In the 1920s the quantity surveying role was initiated, and extended with the publication of the first edition of the Standard Method of Measurement. Building companies were growing larger and beginning to take on work nationally, in addition to local commitments. The first system-built houses appeared in the late 1920s, demonstrating a desire to move away from traditional methods of construction. No fines concrete, steel frames and timber frames were introduced. Concrete was being used extensively in columns, beams and floors to commercial buildings. Progress was being made in the science of concrete in European countries and the UK benefited from the dissemination of these advances. The industry took on a further, perhaps unwanted, function, that of an economic regulator. As more government money was directed to building houses for rent, it took an increasing proportion of the public expenditure budget. Local government built the houses with this money and from local revenue such as rates. During the 1930s, unemployment was high and to try and reduce the number of people out of work public money was spent on building houses, schools and ancillary services. The dependence on government sponsored building work increased, until by the mid-1970s half of all money spent on building was funded directly or indirectly from public sources.

There was not a significant amount of development between the wars, with the exception of a small number of system-built estates. The majority of buildings did have electrical power, and appliances such as cookers and fires were common. The number of power points were few, restricted in the main to the kitchen and living room. Cavity wall construction was used on some houses in the 1930s, but it was not until after the 1939–45 war that this form of construction was used on all brick-built dwellings. Open fires were still the main form of space heating, although gas and electric fires were increasing in popularity. Hot water radiators were confined to large commercial or public buildings. Steel enabled the average building to have large-span roofs which could be covered in corrugated asbestos sheeting. Domestic buildings had clay and slate as alternative roof coverings. The role of technology was to produce a simple functional building providing solid comfort with no pretentions, not to advance dramatically the methods used.

At the end of the second world war the public's attention was focused on quantity and quality of housing and other buildings. This was due partly to the damage suffered from the bombing and partly to a raising of expectations among the population of the UK. This desire to improve both the number and quality of buildings was seen all over Europe, both east and west. Initially, quality had to be sacrified as materials were scarce and costly and the demand for dwellings took all the available materials. Sub-standard houses were demolished and new ones built, along with the replacement of bombed buildings. Demographic changes – the increase in population and the growing need for homes for young couples – also created a need for more houses. To meet this massive demand upwards of 300,000 dwellings a year were to be built in the UK. This target could not be met by traditional building techniques as the numbers of skilled craftsmen and materials were inadequate. Thus, precast concrete panel technology was employed to build fast and high. Designers were concerned to improve the environment: town centres with back-to-back houses were now considered undesirable: space was needed between the buildings, with gardens front and back. To achieve the desired density of people to an area some would be housed in multi-storey blocks of flats. These would be built to the highest standards, with central heating, constant hot water as required, power points for electrical appliances, kitchens with fitted cupboards and worktops, modern sanitary fittings and lifts for access. The majority of this housing was built by local authorities for rent. Private house building relied upon traditional members of construction. It was considered that the level of knowledge of building in concrete was of a high enough standard to meet the long-term performance requirements. Unfortunately, this assumption has proved to be false. With hindsight, criticism and condemnation can be made of mistakes both in design and construction, but this does not take into consideration the mood of the times. Society was concerned about the numbers and quality of housing and was demanding, via the ballot box, that new standards should be met. The construction industry had to meet that demand but, as is now known, it was ill-equipped to provide the right level of quality. Technology did not fulfil its role, perhaps owing to the pressures imposed upon it.

Concrete was used not solely for domestic dwellings but also, extensively, for structural walls and frames to commercial buildings. Examples are the South Bank arts complex and the Centre

Point office block in London. Whilst failures such as weather penetration and cracking/spalling are not common on these structures when compared to housing units, their appearance has been adversely affected by staining. Generally, the public sees concrete as an unfriendly material which creates harsh environments. It is strong and durable when mixed and placed correctly but if not done to the highest standards will rapidly deteriorate. It was initially thought that concrete technology could solve many problems, allowing fast construction of buildings, with large clear functional areas and to a great height. An example of this is the bus garage at Stockwell, London. However, only a few concrete buildings combined the walls, columns and roofs as an integral structure, because formwork construction was complicated and costly. Space-frame roofs, which superseded them, are an example of the technology of building having to provide a follow-up solution to another technological advance: the so-called jumbo jet with its large wingspan required a canopy under which maintenance could be carried out, and a professor of engineering invented the spaceframe structure to house these machines. The use of computers has enabled complicated calculations to be solved easily and has also provided simulations of the structure in use. Designers can see three-dimensional graphic models of their proposals and can create simulations of their internal and external environmental performance.

The functions of building technology in the present industrial and leisure society are as diverse as the demands they are trying to meet. They can be described as:

1. providing in buildings high levels of internal comfort which can be accurately controlled, such as space heating, efficient lighting, adequate power, functional and good-looking appliances, air conditioning, sound and thermal insulation;
2. creating pleasing, maintenance-free external envelopes;
3. supplying functional buildings within which commercial and manufacturing activities can be carried out efficiently. In many cases the structures are designed and constructed to satisfy the demands of the processes which they accommodate;
4. providing suitable environments in which modern arts and sports activities can be enjoyed.

Fig. 1.4 briefly illustrates and summarises the connections between the types of buildings, their prime functions in society and the organisational characteristics of the construction industry.

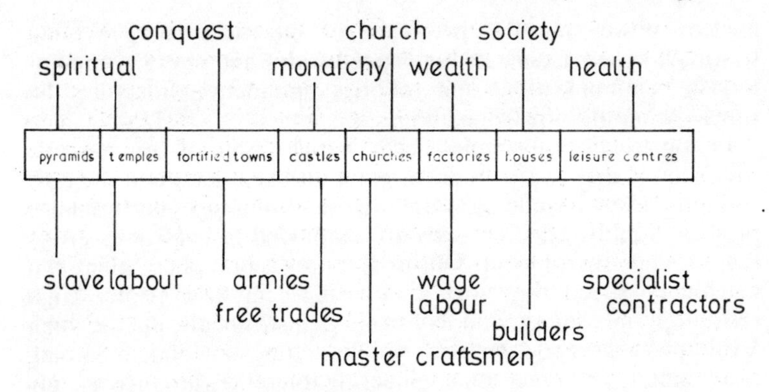

1.4 Types of buildings: roles and producers

FORMS OF TECHNOLOGY

Technology as described in the previous section was connected with the power of the state and the Church and, latterly, with the aspirations of a nation's people. As practised by the western industrialised countries it has developed over many years and undergone many changes. Throughout the world it can be found in many forms, up to the most advanced level of high technology. An example of this is the heat recovery system, in which a central atrium funnels air to its apex where it is recovered and directed down and within the external envelope to be circulated through the building again; it depends on sophisticated air circulation and control instruments. Here, the whole building is designed to maximise the floor area and to house the services on its perimeter, most of these being visible from the outside, such as the prefabricated toilet pods.

Schumacher has suggested that technology should be matched to the economic and social structures of countries, especially those with minimal resources in materials, skills and money. This he calls appropriate technology. It may or may not utilise high technology but is geared to the ability of the people to carry out the processes for themselves. Therefore, a solution might be to build in simple solar-dried clay bricks, effectively reducing the scale of building to meet the low-strength characteristics of the bricks.

The various types of technology are described below.

1. High technology

The majority of industrialised nations base their economic and social development on the use of high technologies. These include:

nuclear power for the provision of electrical energy; rapid transport systems, such as high speed trains and aircraft; computers in communications and robotics; interactive video for the transmission of knowledge and ideas.

In construction, examples occur in the design of components: for example glass which reduces glare and heat loss, and the now common use of cladding panels made from high performance plastics. Sophisticated services are employed to control carefully the internal environment. During the production phase plant and machinery speed the work and reduce physical effort. High resolution measuring and controlling instruments enable high levels of accuracy to be achieved in setting out and in joining components, and precision engineering tolerances are increasingly used.

2. Low technology

This form of technology endeavours to bring back into use that which would otherwise be discarded as waste. A common example is the use of broken brick as hardcore fill below solid floors. Refurbishment and renovation of an existing structure is a basic low technology, even though it may use materials and components produced in a high technology manner. The existing structures are recycled to meet contemporary standards and requirments.

Those technologies used in third world countries can be described as low. The machinery employed is simple and there is a heavy reliance on human labour. Buildings are constructed using local materials and do not contain complex services.

3. Appropriate technology

Schumacher was the first to focus attention on this form of technology. He said that high technology can work in those nations wealthy in natural resources and skills, but those with limited resources should not try to adopt technologies which have been successful under different economic and social circumstances. There are too many examples of developing nations adopting technologies which have failed to work effectively. Take, for example, the new city of Brasilia – beautiful buildings with high levels of services set in landscaped surroundings but not used, or only inhabited by people who are not able to utilise the facilities provided and cannot pay for their upkeep.

According to Schumacher appropriate technology should improve the industrial base of a developing nation by:

(a) creating workplaces where people already live;
(b) making those workplaces cheap to build;
(c) for the actual building and the industrial process which it accommodates, utilising simple methods, skills and materials;
(d) using materials which are indigenous to the area and are mainly for local consumption.

4. Convivial technology

This form of technology is founded in a social system which puts the tools and techniques into the hands of everyone. Instead of relying on large, separate, centralised, specialist centres of production and commerce each community would be responsible for meeting all its own needs, with greater autonomy residing in decisions at local level. Convivial technology rejects the principles of the economics of scale and extensive trade between communities.

5. Liberatory technology

This concept is based on the idea that man can be freed by machines from the burden of work. Minimal labour would be required and the number of hours worked would be drastically reduced, thereby liberating people to enjoy other pursuits. Would this be the utopian life? Some aspects of high technology are already reducing the burden of physical and mental effort, but not the worry of organising and controlling activities. A major change in the ways of allocating work would need to be instigated so that all sections of the population able to contribute to the production of wealth and services could do so. The problems of high unemployment manifest in the 1980s show that the equitable distribution of work is difficult to achieve.

If liberatory technology is pursued to its limits, so that work is carried out solely by machines, will mankind be dominated by the demands of the machines? Could an 'intelligent' machine rule the world?

6. Alternative technology

Most industrialised and developing countries are moving along a path which leads to even higher levels of technology. More nuclear power stations are being built; more factories being set up; more robots being used in manufacturing processes. The latter two demand a large amount of energy to deliver the goods. Much of this energy is lost through poor levels of thermal insulation in buildings. The Friends of the Earth Group have put forward an alternative to the construction of new nuclear power stations in the UK. They calculated that by the turn of the twentieth century two new stations will be required to supply the extra power needed. At 1986 values these will cost at least £1.1 billion each. Approximately half the energy produced in the UK goes into warming buildings. By increasing the thermal insulation of existing buildings the power output of these two power stations would be saved. Jobs in construction would of course be created by the construction of the power stations – a total of 11,000 over five years would be realistic. However, to bring buildings up to a high standard of thermal insulation would require much higher labour input, calculated at 40,000 jobs over a period of ten years. The cost to the nation for this programme of insulation would be the equivalent of the construction costs for only one nuclear power station.

There are wider issues that need to be considered in this debate, such as national politics, the future development of high technology, relationships with other countries, armaments, safety, nuclear waste disposal and alternative energy sources for the distant future. Technology does offer alternatives and these should be considered before any decisions are finalised. As knowledge develops it enables further aspects of the problem to be analysed and brought into the equation; far from limiting the options available, present day development allows consideration of a widening range of options.

SUMMARY

The role of building technology over the years has been outlined, from prehistoric times to the present. It has developed from the basic provision of shelter, through buildings representing the power of the state or religion, and those primarily concerned with the generation of wealth, to the current mix (in industrialised nations) of all these functions plus those of entertainment and leisure.

Construction methods, processes and types of organisation have also changed and developed. Building and building craftsmen were highly thought of by society during the middle ages and perhaps reached their highest standing just prior to the Industrial Revolution. Since then, building has been seen as only a means to an end, for example by providing functional structures for the manufacture of wealth-creating goods.

Architecture is still held in relatively high esteem but is currently undergoing criticism, particularly with respect to the role of architects in the construction process and the content of their education and training.

Although individual master builders, as respected and influential members in society, have lost their status, the credibility, professionalism and commercial integrity of building organisations has risen. The services offered by project management teams exemplify this.

Technologies must be matched to the context which they service, and a number of different types have been briefly described. At present all these types are being employed throughout the world, with variations even in the UK – for example, low technology in refurbishment; liberatory technology in computer-controlled warehouses; convivial technology in the involvement of local communities in providing dwellings etc. in urban environments. All of these should be used as appropriate to the needs of clients and of society.

QUESTIONS

1. Discuss the role of building technologists in an industrialised society.
2. Describe a construction project, decide what type of technology is appropriate for it, and explain why.
3. Discuss the merits or demerits of one technology type for the majority of buildings in a society.
4. Present a scenario, based on historical evidence and current trends, for the state of development of building technology 25 years from now in your country.

Part Two
SITE PROCESSES

4. Site Conditions

The site and its surroundings have a major impact on the type of building to be erected and the methods of organising the contruction work. The site can determine the technologies, especially where further constraints are imposed, such as poor soil conditions or restricted access. These can be considered under the headings of *geotechnical* factors and *environmental* factors.

GEOTECHNICAL FACTORS

The structural stability of a building is founded on the characteristics of the soil. A full knowledge of the soil is necessary to undertake foundation design. It is also necessary for the builder to be aware of the expected site conditions so that appropriate site technologies can be planned, costed and used. The common forms of site investigation techniques and the information required to ascertain soil characteristics are listed below. As the concern here is with the implications of the builder's knowledge and its effect on the building technology, this list is to be seen as an aide-memoire only.

Soil Investigation Techniques

Trial pit/hole. Allows a visual inspection of the soil and its strata; disturbed samples can be obtained.
Core sample. An undisturbed core sample can be taken from a great depth. The soil from this sample can be tested in a laboratory.
Seismic testing. This will indicate the overall nature and thickness of the various soil strata.
Core penetrometer. Gives an indication of the soil's consistency and characteristics, with reference to its load-bearing capacities.

Nature of the soil

Tests can be carried out on the soil to ascertain the following, depending whether it is rock, cohesive or non-cohesive. The data that can be obtained are:

bearing capacity
angle of repose/slip
ratio of water to pores etc
plasticity
particle size
proportion of other types of soils.

The science of soils is still developing and the ability fully to predict behaviour has not yet been achieved. In the majority of investigations and analyses the ground is likely to perform within certain parameters based on the information obtained.

Borehole logs are the main method of communicating the information regarding the strata of soil. The additional information on soil analysis is usually reported separately. The borehole log may only delineate the main levels of strata and identify each soil type, together with information on the water table level. This will be sufficient for most lightweight buildings up to three storeys in height. Where the ground conditions are known to be poor, a full analysis of the soil should be sought. Any structure above three storeys should be designed on the basis of a full soil analysis. The number of trial pits and core samples is important. Unfortunately, research has shown that there is commonly insufficient information regarding the soil conditions over the whole site (National Economic Development Council, *Faster Building for Industry*, 1984). Although site investigations are undertaken their scope is extremely limited. The nature of the soil can change even within the confines of a small green field site. On town sites, where buildings and services already exist, the need for soil and ground condition investigations to be carried out thoroughly is vital. Sites which are being redeveloped will have old foundations and services. There is a need to map the services and other structures, not only to inform the builder of their presence and to indicate the extent of work required for their removal or repositioning, but also as a matter of safety. Too many serious and fatal accidents occur on site through people cutting through electricity cables or gas pipes. In some cases the services were not shown on any drawing provided by the public utilities; in others the architect or builder had not checked to see if there were any on the site. Lack of knowledge with respect to ground condition will, at least, cause disruption to the design and building, as measures will have to be undertaken to cope with the problems. Extra costs may be involved, some of which might have been avoidable if a thorough site investigation had been carried out initially. On sites previously used or built upon it is necessary not only to take soil samples but also to determine the extent of underground structures and services. The state of the services, too, must be ascertained, i.e. whether they are live or dead.

Generally, site investigations are carried out by specialists in geotechnics. They will carry out the investigation to the client's/ architect's instructions. In other words, the scope of the investigation will depend upon the instructions given. Where obstacles have not been found it is not usually the fault of the investigating organisation as they should have been told how many bore holes etc. would be required. Of course, the more boreholes and tests the greater the cost – this being the governing factor in limiting the scope of investigations. To carry out a full investigation is costly and has to be weighed against the cost of the proposed structure. Large buildings in known uncertain ground will necessarily require a full investigation. Regrettably, there are cases when this has not been carried out and the soil is found to differ on various parts of the site. A prestigious building in the City of London recently had these problems. The soil when excavated in one section of the site was not as indicated by bore holes close by. A major redesign of the foundations was required, causing a delay resulting in extra costs and a slippage in the programme of two months.

Interpretation of soil investigations is the responsibility of the designer and builder. Even though full details of foundations etc. may be supplied, based on criteria found in the soil reports, the builder must satisfy himself that the designs are appropriate. If not satisfied, then further investigations, analysis or redesign should be called for. Additionally, interpretation is needed for the design and planning of any construction activities associated with the work in the ground. The builder may need to carry out further investigations to obtain information for the design of temporary works, such as support for the sides of excavation. These must integrate with the general construction process of the building.

Whatever the job or the site it is incumbent upon the builder to understand the soil and the site conditions and carry out the construction work with respect to the known data and in a safe manner.

Effects on construction methods

The poorer the site and soil conditions the greater will be the effect on design and production of the building. Let us consider the problem of a high water table.

In the first place it means that the soil is affected by water, which has consequences for its characteristics and bearing capacity. Therefore, the design of the foundations (and/or below-ground structures) should take full account of the soil. During site production the water will have to be controlled. The choice of controlling technique will be dependent upon the following factors:

size/area of site (space for installation of equipment)
depth of water table
depth of proposed foundations/groundworks
proximity of adjoining buildings
effect of water control on adjoining land
type of foundations, requirement for working space and access to excavations
time available to carry out work
means of dealing with controlled/extracted water, and its disposal.

Common groundwater control systems are:

sumps and pumps
dewatering with well points
freezing of the ground
electro-osmosis
piezometers
cut-off walls.

According to the extent of the factors above – and the economic factors – a choice will be made. A few of the implications of these factors will now be discussed.

The size and area of the site may make the use of extensive equipment, such as well points and pipes, impractical. Their presence may impede the efficient construction of the building. A large site may need to be divided into sections and phased, otherwise the water control system will be large and expensive. If large volumes of water are being extracted its disposal may be a problem. If the water table is high then some form of control needs to be carried out immediately. Again, the volume of water may cause problems. The deeper into the groundwater any structures go the more acute the problems; one is the increase in the pressure on the sides of the excavations. The use of cofferdams will have to be considered. Extracting groundwater may have a detrimental effect on the adjoining buildings by disturbing the balance of the soil under their foundations: this could apply to adjoining land. Cut-off walls might have to be placed to restrict the water extraction to the confines of the new building site. Even where there is no effect on adjoining property, a cut-off wall might be necessary to avoid a continuous inflow of water. If cut-off walls are not used freezing might be an alternative, although a disadvantage with this method is that the ground takes a long time to reach a satisfactory frozen condition; excavation will also be harder in frozen ground.

The type of foundation required may necessitate an allowance for working space or access. In this case cut-off walls with local control of the water may be the answer. If time is at a premium, the choice may centre on that method which can most quickly deal with the water, with cost being only a secondary consideration. Some local authorities may impose restrictions on the disposal of water because of the inadequacies of their sewage systems. Pipe sizes or treatment plants might not be able to cope with the increase in groundwater.

Where basements are to be constructed in ground which is poor, whether because of the nature of the soil or ground conditions, there is a greater opportunity to employ coping strategies. Basements require wholesale excavation over relatively large areas, with the need to provide retaining walls. These can be a means of controlling water, of giving support to adjoining property and acting as the foundation to the new building. In the case of isolated foundations in poor conditions the works necessary to instal these can be comparatively complicated and lengthy.

A problem which is likely to become more common in the future is that of old foundations or structures on a site. Assuming that their presence, and extent, are known to architect and builder, how are they dealt with? There are a number of factors which will influence the course of action:

type of new building on site
structure of the building (framed, load-bearing walls etc.)
type of foundation (related to soil and structure)
extent and depth of underground structures
material of underground structures (brick, concrete etc.)
relationship of old to new structures
restrictions or constraints on their removal, e.g. proximity to property.

The type of building also will have some influence: its height and area, together with its function. For example, a factory whose floor has to take heavy loadings at any point will need to be founded on ground which does not have obstacles which could create fulcrums under the slab. If the building is framed it may be possible to design the foundations to miss the existing structures, thereby saving time and effort in their removal. In the case of continuous beams or closely packed piles this might not be possible. If the obstructions are left, what will be the effect on the distribution of the building's loads to the ground? It may be that the relatively undisturbed soil may not be strong enough to carry the loads, or that its strength is incompatible with the madeup ground. Both existing soil and old structures will need to be levelled to a suitable load-bearing stratum. If the old structures are very deep they may be difficult to extract, so the new foundations will have to be designed around them. If they cover the site it could be impractical to remove any but those in the way of new work. The material may cause difficulties in extraction, particularly in the case of reinforced concrete. Breaking out and disposal will then require special treatment. It may not be possible to remove some obstructions owing to their proximity to, or support for, adjoining property, or their removal could involve major new support works.

Two cases have been considered above, namely, the water table and old foundations, with the effect they might have on technologies for the design and construction of buildings. The wider environmental factors which must also be considered, in conjunction with the geotechnical, are now presented.

ENVIRONMENTAL FACTORS

The environmental factors affecting building technology can be divided into three categories: general climate; local area; site.

General climate

Weather characteristics will vary from country to country, and from region to region within a country. Building work can be directly affected by climatic conditions. Bricks can be impossible to lay in heavy rain as the mortar will be washed away. Hot dry air may necessitate the soaking of bricks to reduce absorption of water from the mortar. Protective measures need to be taken in cold or hot weather when laying and curing concrete. In the English Lake District, for example, the rainfall is relatively high which means that rain is likely to fall on more days and at a higher rate than in other parts of the country. This must be noted when programming and devising work methods – temporary covers over the whole building might be one answer. Local wind conditions might create problems, extra care being needed in the support of partially completed buildings. In the UK there are particular problems relating to the climate, in that the range between hot and cold, calm and stormy, wet and dry, can vary within one region within a short space of time. The forecasting of these conditions may not be accurate. Consequently, measures, such as enclosing a building in a temporary protective envelope, could be viable in one winter (as regards cost benefit criteria) and not in another. The decision to carry out extensive protective measures, or not, cannot be based on definitive data, since there is a large element of uncertainty and risk. Hopefully in the future it will be possible to produce better short- and long-term weather forecasts.

An aspect which is just beginning to receive attention is that of the climate generated by the building itself, especially with respect to winds. While this manifests itself mainly after completion, the effects can be noticeable during construction, especially when the basic structure is complete. Although not fully substantiated there is reason to believe that wind patterns created by a building can lead to problems with the building itself. One such example is the Hancock Building in Boston, USA, where during construction the large sheets of cladding glazing 'popped' out from their façade. Because of legal problems and a case which was settled out of court, the evidence of experts presented by all parties – engineers, architects, builders, glazing suppliers, independent experts – is not available to the public. Consequently, it is not known for sure what caused the glass to fracture and fall out. However, some of the remedial works put in hand were designed to prevent the movement of the building in the wind. One expert said that owing to the orientation of the building unusual wind conditions and pressures were occurring around the lower storeys. These were found to be greater here than at the top of the building.

Local area

The local area is the immediate surroundings of the building site. How can it affect organisation and methods on the site? In the case of a green field site one concern should be access to the site. Are the local roads wide enough? Are they strong enough to take heavy loads? Are there any low bridges or other possible obstructions? In considering a town site the off-loading facilities might cause problems. Will the vehicles have to park in the road? What disruption might be caused to traffic movement? Will deliveries be restricted to certain hours? Because of local restrictions, lorries might have to be smaller than usual, creating extra transport costs. The size of components may also be restricted as a result, thereby causing a rethink in the building's technology. Appropriate permissions may be required in order to allow temporary street parking, or to prevent other cars parking in access roads. A road may have to be closed or made one-way. The environment of the local area must be taken into account to see if it will unduly influence the technologies proposed for the design and construction of the building.

On site

The site has the greatest effect on the building's technology. Although difficulties can be predicted many problems arise during the work. For example, a particular activity may cause an excessive amount of dust, and complaints from adjacent occupants will make the builder adopt an alternative methodology to reduce the nuisance. Maybe different tools or equipment will have to be used, or the dust damped down with water, or protective covers erected. An awareness that building activities can generate an unacceptable atmosphere to those people off-site should be fostered. Noise, dust, dirt and the general comings and goings can cause a nuisance. Statutory legislation and the laws of tort provide a framework within which acceptable levels of nuisance can be ascertained. The need to reduce the noise of machinery, particularly pile-driving plant, is obvious, but an accumulation of small tools' noise can be as much of a nuisance. Problems can arise from the intrusion of plant and machinery into other properties, for example, the jib of a tower crane protruding over adjoining property, or scaffolding sited close to an adjacent property causing an obstruction to light. It may be necessary to inspect and assess the condition of adjoining property before building work starts. Reports are then prepared illustrating the state of repair etc. and if there is subsequently a dispute over damage possibly caused by building work then the reports should show whether it was or not. The proximity of other property to the site may allow working from one face only. If the building rises above those surrounding, then cantilevered scaffolding could be used.

It must not be forgotten that on-site environmental problems exist on green field sites, especially where parts of the site are handed over and occupied while construction is still going on. A private housing development is a case in point. Houses are completed and sold, with the owners moving in whilst other houses are being built around them. Constraints will be placed upon the builder: movement of materials around the site will be restricted to roads, avoiding the gardens of houses; materials cannot be stored just anywhere; and operatives will have to be circumspect in their behaviour.

Each of the environmental factors – climatic, local and on-site – will oblige the design and building activities to take account of their effect. Information will have to be gathered and incorporated into the planning and programming of construction. Many of them can be foreseen and their effect on production predicted. To ignore them will lead to inefficient working and inappropriate technologies, with probable loss of productivity, quality and money.

TEMPORARY FACILITIES AND SERVICES

Nearly every building activity requires some support facilities or services. The extent of these facilities will depend upon the following:

size of work task
technological complexity of the work
length of time required to complete the work
number of people required.

The size of the work task may demand a large number of services. One example is a large housing development built in a phased sequence. As the work spreads out the temporary services will have to precede the activity. As distances between facilities, such as toilets, become longer, it will be necessary to provide more toilets to reduce inconvenience and time spent walking.

A second example could be a multi-storey building. As the work goes up so will the services rise along with the floors. Again, intermediate toilet facilities should be provided to reduce the time wasted in travelling.

The technological complexity of the job may demand a number of types of powered plant and equipment. Elaborate electricity supply circuits will need to be installed. High levels of precision engineering tolerances and accuracies may necessitate the use of sophisticated measuring and placing instrumentation. The assembly of the technology may be governed by how easy it is to ensure its correct placement. In other words the provision of measuring instruments and their placement on site can influence the sequence of assembly operations. If electricity is required it may be best to use generators, rather than get a temporary connection to a mains supply.

If the job is to be completed within a few weeks the provision of full health and welfare facilities is not necessary. The number of people on the job will determine directly the provision for safety, and of such facilities. Reference must be made to the Safety, Health and Welfare Act 1974 and the Construction Regulations to ascertain the particular requirements: the larger the number of people working on the site the greater must be the provision of services (four gangs of bricklayers will require more stand pipes than one gang).

The main facilities considered here are: accommodation; services; and security. The important aspect of safety is dealt with in the next chapter, on aids for production.

Accommodation

The type and scope of accommodation is primarily covered by the requirements of the legislation mentioned above and by the demands of the site. If there is a large number of office-based staff then suitable accommodation will be needed. This will have to comply with the Offices, Shops and Railway Premises Act 1963. The number of office staff will be determined by the demands of the job. Some, such as general foremen, may require a desk, chair and minimal storage space for papers as their work is mainly on site. Others, such as quantity surveyors, would require more storage space. A large room for meetings is necessary: this may double as the site manager's office. Other facilities might include toilets and a place for making drinks and preparing food. Provision may need to be made for a computer, and this may cause problems as the general atmosphere on a building site can be harmful to the computer and its disks. Dust is a major hazard here; also, most site offices are not built to a high standard of thermal conductivity and it is difficult to maintain even temperatures – extremes of temperature can disrupt the efficient working of a computer. Perhaps in future greater attention will be paid to these matters as it is inevitable that micro computers will be used more frequently on site, especially with large complex projects.

Other accommodation will be required for the site operatives, such as canteen/mess hut, drying room, toilet and washing facilities. In some cases specialist contractors might want workshops in addition to their stores.

There has been a tendency for the number of site staff based in site offices to increase, and for the numbers of operatives to decrease slightly. As the construction process becomes more complicated in its organisation of many specialists and subcontractors, and a corresponding concern has arisen with productivity and costs, the staff required to control these has increased. In addition some of the consultants, especially engineers, spend much time on site, and they will require accommodation. Consequently, on some sites there is a veritable village of accommodation.

The quality of office provision is largely at the choice of the builder. In some situations the client may request a high standard so that prospective users/lessees or purchasers will be impressed. It is now common to use proprietary purpose-made units, supported on adjustable jacks. These can be fitted out to any standard and type of accommodation. They may be hired or bought. When finished with on site they can be lifted on to a lorry and transported to the next destination. They can be linked together to form suites or stacked one on another.

Adequate, comfortable and clean accommodation for all on the site can be a useful aid to the achievement of high productivity. It will set a standard of quality for the work itself.

Services

The basic services required on site are: water; electricity; telephone; and, sometimes, gas. It is necessary to notify and order the connections to the mains from the relevant authorities.

The extent and type of services required will depend on the four factors described at the beginning of this section. The use of electricity has increased to cope with the increase in the number of small power tools used on sites. Purpose-made site supply cables can now be obtained, together with portable distribution boxes fully protected against accidental damage. Circuits are required for lighting and for tools and equipment. The cable runs and distribution points should be planned carefully to avoid a confusion of cables: it should not be left to the operatives to set out the cable runs, and all such work must be undertaken by a qualified electrician. The provision and maintenance of this service could be a full-time job. The siting of cables and boxes should not hinder the production processes. For example, they should not run over a floor area which is to be screeded: they will have to be suspended from walls or ceiling, with the distribution boxes on purpose-made brackets.

Carelessness in the use of water can have disastrous effects on partially constructed buildings. There is many a recorded incident where water has been left running and spreading to wooden floors which have then swelled and distorted.

Remote sites may have difficulty in obtaining a telephone connection, though their remoteness makes their need for it that much greater. Even relatively small sites with a life of only a few months will need a telephone. Having a computer on site and a telephone will enable communications to be made directly to sister computers anywhere in the country. This can make information flow much easier and provide data quickly to aid efficient decision making.

Some sites use a system of radio-controlled communications, for which a licence is required, within their boundaries. On a large site with a great deal of movement of people and equipment it can be a boon for locating of people.

The use of gas on site is limited and it is invariably supplied in cylinders. Care needs to be taken both in using it and storing it. Any appliances must be properly vented. It can be employed for cooking and space heating, either in office accommodation or on the site within the new work, but its use in the building must be carefully controlled so that the fabric is not affected by a too quick drying out process or by excessive humidity which could cause condensation problems.

Security

It is important for construction sites to have adequate security provision. This has to work in two ways: one, to prevent loss of materials etc. during working hours by operatives; two, to prevent entry of people wishing to steal from or vandalise the site. Research has shown that most losses from sites occur during working hours. The control of this relies both on the site's fencing and on the diligence of the site staff in monitoring activities and materials. The prime function of fencing is to keep people out. Town sites usually have close boarding or sheets. There may be provision of inspection openings for the public to view the works or to allow for setting out instrument sightings from outside the site. The choice of fencing should reflect the degree of security required and the efficient operation of the site technologies. It will be of little use having a high security fence which impedes access to the site or prevents the optimum placement of temporary works.

Open field sites usually operate to a system whereby all materials, plant, equipment and accommodation are situated together within a fenced compound. The fence may be open mesh or solid sheet. Where it is in full view of passers-by it is better to have open mesh and floodlight the area at night – any intruders can then be seen easily. It also acts as a deterrent, as no intruder *wants* to be seen, while anything can be happening behind a solid fence! In winter the floodlights will be useful in prolonging the working day beyond daylight hours.

The location of a site will affect the measures adopted for its security. Town centre sites adjacent to housing estates can be damaged by vandalism. Those in less populated areas are susceptible to the theft of materials and components. In both cases there is ingress to the site, but with different results. Whatever security measures are adopted they must be maintained. For example, a leisure centre contract had a large tidy compound located adjacent to it. The site offices and stores provided one boundary and the others were wire mesh fencing topped with three strands of barbed wire supported on concrete posts – an adequate and effective fence. Unfortunately, a hole was made in the fence. This may have been unavoidable, but what was avoidable was the failure to repair it until two months later. The whole security system was during that time flawed and, therefore, useless. The cost of repair would have been as nothing compared to the potential losses.

A passive security system can be enhanced by an active element, i.e. patrols or a resident night-watchman. If either is used they should be able to check the passive system easily.

In this review of temporary facilities and services some idea has been given as to where they are influenced by building technology. For example, accommodation must service the needs of the workforce, which in its turn is determined by the building's technology; in the provision of services the use of electricity is increasing (aspects of safety are all-important here, the first rule being to transform down to 110 volts, if taking the supply off the mains); and the telephone has become an essential link in the communication system.

CLASSIFICATION OF THE SITE

The better to describe the variety of sites, they will be classified into three types. The classification is made on the basis of the free space available to the builder for his operations, accommodation and storage:

1. open field
2. restricted
3. very restricted

1. Open field

The problems encountered on a green field site not previously built on, with plenty of space available for building operations, are relatively few. Trees, shrubs, etc. may need to be removed and the ground stabilised. Site roads, temporary or permanent, should be laid at commencement to avoid the possibility of transport cutting up the ground or getting bogged down. A possibility that could arise is that with plenty of space around the site, materials will end up scattered over a wide area. They may not necessarily be close to the point of use: the work could progress past them. When located at the site boundaries they will be susceptible to loss by theft or damage by vandals. Valuable materials should be placed within a high security compound, where plant and machinery can also be parked overnight.

There should be few problems in site clearance, other than the consolidation of ground after the removal of large trees.

2. Restricted site

With a restricted site the building takes up the majority of the space, or there is some other obstruction. It might be located in an open field or in a built-up area. Considerable care and attention will need to be given to the positioning of site accommodation and to the storage and movement of materials. Construction will need to take into consideration these problems by providing clearly defined movement corridors and storage areas. It is possible that alternative positions will be available, and the choice of the optimum is critical to ensure efficient working.

3. Very restricted site

A typical very restricted site will be located in an urban area, bounded by buildings and with the new work occupying the whole area. With virtually no room on the site the sequence of operations is dependent upon the rate of flow and placement of materials and components.There will be little or no choice in the positioning of accommodation or storage areas (these may have to be off site). A high degree of control over site operations will be needed.

The layout of each site type is considered later.

PREPARATION OF THE SITE: DEMOLITION

Many sites contain existing buildings or structures: these need to be removed before new buildings can be erected. Removal may be the responsibility of the builder, or it may be under a separate contract with a demolition specialist. It could be that some demolition work has already been carried out before the builder is appointed, or the use of the site decided. Further work may then be required to complete demolition or clear up the site. Demolition work can be carried out to the following final stages:

complete demolition and removal of the structures to below ground foundations, with any required support to adjacent property or ground;

demolition of structure and removal to ground floor slab, with any required support to adjoining property;

demolition and removal of ground floor slab, but not any groundworks.

The demolition contract can have two options: one, that all materials belong to the demolition contractor and he has responsibility for their clearance off site; two, the materials are the property of the site owner who retains responsibility for their clearance. Under this latter option the demolition contractor will estimate for the work involved in carrying out the demolition without off-setting against these costs the sale of the waste materials. The estimate will therefore be higher than in the case of the first option.

It is not unknown for a builder to take over a site and find it covered in demolition waste and other rubbish awaiting disposal. The code of practice on demolition (CP 94/1971) sets out the methods for good practice on site. The majority of recommended methods are based on safe working principles and on protecting third parties and adjacent buildings. There is little doubt that demolition work is more hazardous than general building, as the accident statistics show. Many accidents are due to inadequate safety precautions and non-compliance with the code of practice.

The builder who has to oversee demolition, whether carried out under a separate contract or let as a sub-contract, should be fully aware of the recommended methods and should insist that they are carried out. This can be enforced by stating that all work is to follow the recommendations of the code of practice.

Demolition and building work are often carried out simultaneously in the early stages of a job. Close co-ordination is required between the two organisations to ensure safe and efficient working. Zones will need to be designated for each to work within and a phased sequence of handovers agreed. Problems are bound to arise regarding the movement of traffic on site as it is likely that waste material will be leaving as new materials arrive. Will traffic jams occur? Who monitors traffic flow? Do checks need to be made on lorries leaving the site? Will the dirt, dust etc. affect the new building work? Will any protective screens, fans etc. have to be provided?

Where temporary support to existing structures is required the system adopted should take into consideration the new building and its method of construction. Under the conditions just described the contractor can work with the demolition contractor in providing the right system for the work in hand. There are many cases where temporary supports are erected which hinder the placement of new works to a building. Of course, it could be that the new building was not designed before the demolition took place, and it is quite likely in that case that temporary structures could interfere with new foundations etc. The position of any temporary structures should thus be considered in the design of the new building. Methods of demolition are influenced by the following factors:

structure of the building (framed or load-bearing walls)
materials in the structure and fabric
height and shape of the building
proximity of adjacent buildings or other restrictions (noise, dirt etc.)
need to save any component or element
need to provide temporary supports
need to work with other builders etc.

For instance, a framed structure can be demolished in two stages, the external cladding and internal walls, followed by the frame, storey by storey. If the frame is concrete it will need to be broken down using pneumatic hammers. A tall building will need special equipment to bring the waste material to ground level. Restrictions may limit site working to particular hours.

Where an element is required for re-use its removal or support will require special care and attention. During demolition the building may become unstable, necessitating the use of temporary supports until that element is taken down.

Demolition, and its aftermath, can cause problems for the builder and a full analysis should be made to ascertain the extent of these problems before embarking on the new work. Note should be taken of the type of contract for demolition, its scope and the factors stated above. This will allow an appropriate technological solution to be provided by the builder.

LAYOUT AND ORGANISATION

The section on classification of the site proposed a system of classifying sites according to the space available to the builders for their services etc. In considering the layout and organisation of a site it will be useful to place the site in question within one of these categories. Once the category has been identified then the feasibility factors can be set within appropriate constraints.

Before planning and selecting the site facilities and layout the following questions, defining some principles, should be asked.

What is the site's classification?

Is the site *open field*, *restricted* or *very restricted*?

What is the main purpose of the layout?

The concerns of the builder in the execution of work need to be rated in order of priority. For instance, if there is a need for a large amount of plant and equipment, then the movement and placement of this becomes paramount. If the job requires a lot of loose materials, such as bricks and blocks, their storage and retrieval becomes the paramount concern. If there are substantial temporary works, then their positioning, maintenance and removal will directly influence the site layout.

How does the building technology affect the layout?

The building's structure and components will determine the layout requirements. For example, a steel frame will need a crane for lifting and placing the beams and space for temporary storage. If the building is not tall a mobile crane will be adequate: this will require a movement corridor around the building. Large components might require temporary storage: how many can be stored? Do they need special protection? Can they be easily moved to their final positions? Is it possible to offload and place in one operation?

What facilities need to be provided?

The welfare facilities and office accommodation will be dependent upon the number of people working on the site. Other services such as water and electricity will be provided to suit the demands of the work, their optimal distribution being an essential requisite for high productivity in construction. Also, the need for facilities such as store rooms and temporary works should be assessed.

Can areas be designated?

Ideally, separate areas should be designated for storage, office accommodation, movement corridors and plant. Once designated each should not be allowed to run into another area without due cause. Storage areas have four dimensions: length, breadth, height and time, the latter being very important. At different stages in the job, storage space can be occupied or free. The ordering of materials should be considered in relation to: the space available; the rate of working; handling requirements; the timing of inclusion in the building. A problem can arise if, owing to careless ordering or use, some materials are left behind which cause obstructions for subsequent deliveries. Offices should be situated together to allow easy access and communication between staff. Their position on site will depend upon the degree of control required over the site acivities, the available space and the required degree of proximity to building.

Movement corridors should be clearly planned and delineated on site, whether they are horizontal or vertical. Any temporary roads should be well maintained and kept free from obstructions. Materials should not be left on roads, even if the intention is only to do so for a short time. Any turnings off the road should be marked and kept free from obstructions. In designating vertical corridors all overhead wires, cables, etc. should be noted and if necessary diverted. Plant and equipment should only overhang adjoining property if authorised to do so. All movement corridors for personnel walking around the site should be kept free from obstructions.

How safe is the layout?

Many of the facilities and measures provided for the layout will be subject to statutory requirements, such as the Health and Safety at Work etc. Act 1974. When these requirements and those of the building's technology have been met, the site layout should be critically analysed with respect to the possibility of hazards to safety. The site activities need to be pictured and overlaid, metaphorically, on the site layout in order to ensure that risks to safety are minimised. This may mean that some aspects of the layout and organisation have to be altered in view of a potential danger.

What procedure is required to monitor site organisation?

Once the site layout has been planned and initiated it will need constant monitoring and updating. The site activities are fluid, and unforeseen circumstances could arise: the layout will then have to be adapted to suit. A management input is required to carry out this task. In the first instance the technologist should determine the site layout relative to the factors listed above. An organisation structure and procedure should then be created in order to meet the objectives set by the building's technology. Regarding the control of movement on site, for example, a structure should be established for the processing of paperwork relating to orders and deliveries and their relationship to the programme and site layout. Some control responsibility should be allocated on site to ensure the correct interpretation of the plan and layout. Thereafter, constant vigilance and maintenance are needed.

How can information be communicated?

Plans, drawings, schedules, diagrams and programmes are the prime means of communication. These may emanate from a source not directly concerned with the site activities, such as a builder's head office or consultant. Their intentions should be conveyed to all those working on site, for example by displaying drawings and diagrams on walls in offices or in the new works. When job tasks are set and allocated any factors relating to the site's layout should be mentioned. It may be that some tasks derive directly from constraints imposed by the layout – for example, moving a material via a designated route through a building. Site organisation is the responsibility of all on the job, but if relevant factors and policies are not effectively disseminated then it will be difficult to ensure compliance.

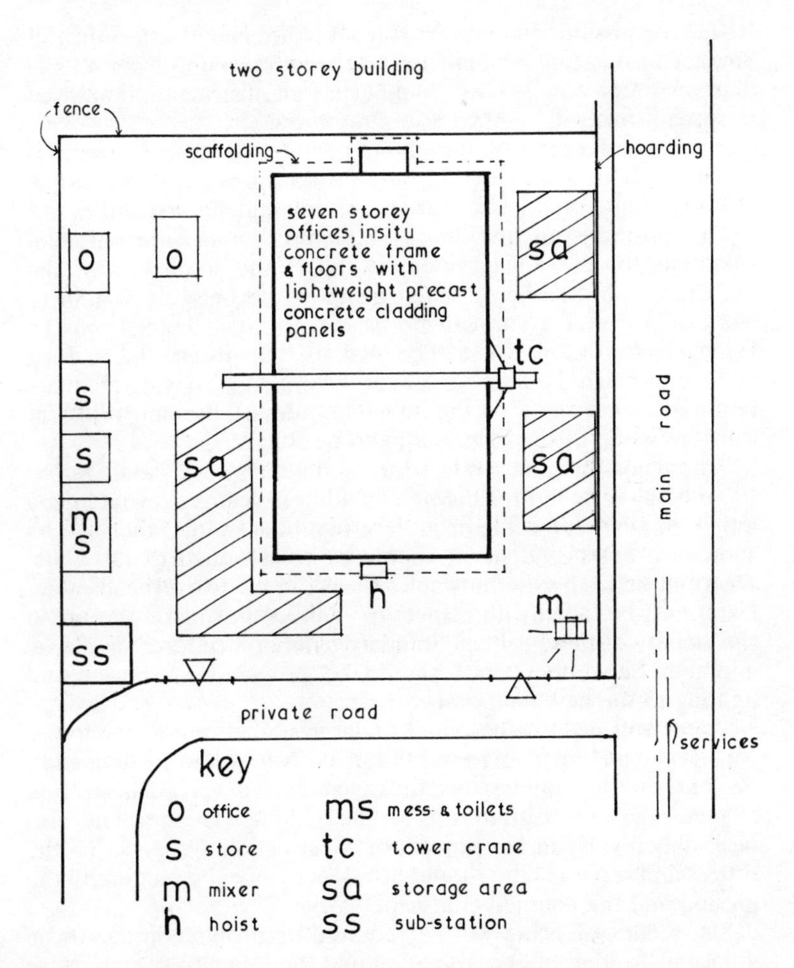

2.1 Site layout

CASE STUDY

In order to give some substance to the factors described above the site layout shown in Fig. 2.1 will be analysed.

The site can be categorised as *restricted*; it has some area around the building for the builder's facilities.

There is some separation of hutting from materials storage and the main access. The huts are grouped to the west side of the site. One can criticise the position of the site office, being so far away from the entrance.

The area set aside for materials storage appears adequate. Materials could be offloaded from lorries parked on the Main

Road (permission to be obtained from police and local authority). Smaller lorries can get onto the site. The position of the mixing plant, with its storage bins, could cause an obstruction if allowed to spread: this will need careful monitoring.

Access and egress is based on a one-way system, which is sensible. There is no specific route to the stores and it is possible that materials could be scattered indiscriminately over this area.

The positions of the hoist and tower crane are suitable. Materials for the hoist can be stored at the south end of the building, with care being taken to allow the one-way system to work. The tower crane can be used to offload direct from the lorries in the Main Road. The span of the jib should be long enough to reach the four corners of the building carrying the mass required at that point. As the jib might oversail adjoining property the permission of the owners should be sought.

Where possible, the services for the building work should utilise those being installed for the new building. Electricity would need either an overhead cable or underground duct within the site. As soon as it enters the site it should be transformed to 110 volts. Distribution to the site huts and offices can be by overhead cable. Care must be taken with respect to vehicles moving on the site in the vicinity of these cables. Standard safety procedures should be adopted. Separate circuits should be provided for power and lighting to the new temporary works.

Water will need to be piped underground to serve standpipes (one at the mixing plant) and the toilets. A foul sewer will need to be laid for the toilets: this to be connected via an inspection chamber to the new drain runs for the building. These will need to be installed early in the process of construction. When siting the offices and services note should be taken of any obstructions in the ground and the planned landscape works.

Site security fencing will be required on all boundaries: those adjacent to the public roads should be 2 metres high close boarding and the other two 2 metre high chain link. The gates could be close boarding or chain link. It is advisable to have floodlights giving illumination at night to deter trespassers.

The layout given in Fig. 2.1 is based on the provision on an actual site. There are many alternative ways of meeting the perceptions and particular needs of each builder. No one layout can be perfectly satisfactory, but, so long as the principles are observed, safety and efficiency should follow.

SUMMARY

In this chapter the general aspects of site conditions have been considered as they relate to the builder. Geotechnical factors are important, not only for the design of the foundations, but for their effect on site methods. A heavy clay soil will become soft and muddy after heavy rain, causing disruption to movement on the ground. The soil will determine the safety measures required for its support during excavations. Building sites are not usually isolated and the local environment will impose constraints affecting site practice, such as limited access. The building will demand certain services for its efficient construction, whether site offices or water. The site should be well prepared and assessed in relation to its ability to cope with the builder's facilities. If buildings were (or are) on the site, the demolition process should be planned with regard to the new building. A site layout should take into account all the above factors.

QUESTIONS

1. Illustrate how the different characteristics of clay soil and gravel soil can affect the choice of excavation plant.
2. Show how different soils can affect the technology of foundation design.
3. Assess which of the three main environmental factors will most influence the technology of a leisure centre building being constructed on the outskirts of an urban area.
4. Plan and describe a process that will result in an efficient site layout.

5. Aids for Production

INTRODUCTION

Site production is primarily dependent on the technological form of the building. Construction ways and means follow upon the demands set by the building's design. In meeting those demands the builder must co-ordinate all relevant and available resources to achieve a satisfactory and safe construction process. The ways of carrying out the work should be analysed and appropriate choices made for each activity and sequence. The means should be ascertained and brought into the sequence at the appropriate times.

A building site is different from a factory production unit. In a factory, work activities are generally static and materials and processes pass through machines or people. During building work people and machines pass through the work activities. In so doing they, literally, have to carry or move the materials and equipment along with them. All access, tools, components, energy, support have to be centred on the work activity in its place. This makes building quite different from the majority of other production processes. The identification and resolution of the most effective production process, related to technological demands, is the concern of this chapter, which will explore how the structural form of the building sets the constraints for production. Factors which have to be considered when transposing raw materials into finished items will be analysed. The process of mechanisation will be examined. Finally, the requirements and effects that temporary works have on the site technologies are discussed in some detail. As can be seen in Fig. 2.2, commencing with the structural form the process is a combination of resources, temporary works and components. The builder seeks to achieve full integration of these factors in order to arrive at the most productive process possible.

STRUCTURAL FORM

The manner in which a building distributes its dead, live and wind loads is determined by its structural form. In the erection process those elements which are designed to carry the loads, such as

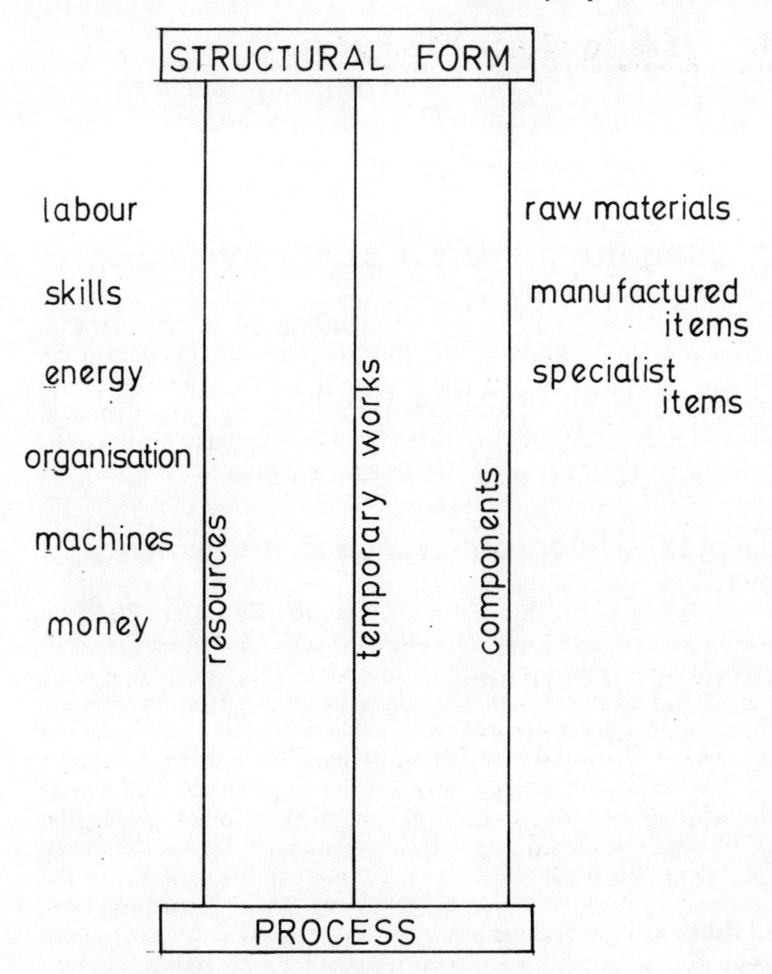

2.2 *Integration of structural form and building process*

floors and external cladding, have to be built first. When these are built the secondary, non-load-bearing elements, can be attached. Therefore, when deciding upon the method of production, the demands of the load-bearing elements must be satisfied. It may well be possible to combine the production of the primary (load-bearing) and secondary elements, for example by combining the floors and supporting beams in one operation.

The principles for production we are attempting to establish are illustrated in Fig. 2.3 and set out below.

1 relate to structural form
(primary element)

2 consider secondary elements

3 integrate primary & secondary elements
↓
produce sequence

4 ascertain resource needs

5 consider safe working aspects

6 plan & integrate resources to elements
↓
produce process

7 monitor process

8 record data & feedback for future
processes

2.3 *Principles for production*

1. Relate to structural form

It has already been noted that any production process needs to be centred on the building's structural form. All decisions spring from the need to erect a load-bearing form and the processes will depend upon: the shape, size and extent of the structural members; the materials with which they are formed; whether formed on site or off site; and the appropriate erection sequence.

2. Consider secondary elements

Following closely on the constraints imposed by the structural form are those generated by the walls, floors, roof and main service functions (e.g. lifts, stairs, heating). These must be considered in conjunction with the prime elements and a process created which will satisfy both sets of constraints. In making decisions based on the structural form the secondary elements cannot be ignored. Can the same plant and machinery be used? What duplicate skills are required?

3. Integrate primary and secondary elements – produce sequence

To achieve a high level of productivity a close integration between the methods of production needs to be maintained. Ways and means of cutting down the number of separate operations should be sought. Method statements should be produced for the major tasks and analysed to find improvements. When satisfactory methods have been devised a sequence schedule should be prepared, and again this should be further analysed to seek improvements.

4. Ascertain resource needs

When the ideal method and sequence has been formulated it is then related to the required resources. What plant? Who follows whom? What aids to production are required? What materials? In effect, a list is produced of those items deemed to be necessary to carry out the activities.

5. Consider safe working aspects

For those activities which are considered to be particularly hazardous a written working method statement should be prepared, highlighting the procedures for safe construction. This can be included in the method statements mentioned in Principle 3 above, but it must be phrased so that all those people on site who are to execute the work can clearly understand the proposed working methods. The written working method statement should include the following items:

- description of activity
- general description of sequence and methods
- any specific reference to the use of safety devices (e.g. harnesses)
- means of safe access and egress from the position of work
- specific details of lifting appliances: their safe working setup and procedures; their position on site to be shown by diagrams
- programme which relates the tasks to the delivery, storage and acceptance of materials
- who is responsible for what.

Once this working method statement has been issued to those carrying out the work (i.e. to all people concerned, from project manager, to labourer, to materials checker, to plant engineer) then there should be no deviation from it without written assent. If a deviation is found to be necessary a variation should be issued by the originator.

The activities which are deemed to be most hazardous, i.e. those having a high accident rate, are listed below:

steel erection
roofing
cladding
painting
demolition
scaffolding
falsework
excavations.

Any construction work involving these activities should be considered a potential high risk, and in the preparation of the tender a plan should be included for safety as well as for the engineering and monetary aspects. The concern for safety should commence at the pre-tender stage and be fully incorporated into the planned sequence of work. To encourage this a five-stage process is proposed.

Stage One: plan for safety

At pre-tender stage a heightened awareness of safety aspects should be built into the planning of the job, especially for those activities listed above. Those people concerned with the preparation of the tender (or in the method of working and budget preparation for private development) should at all times consider the aspects of safety as an essential ingredient of the proposed sequence and costs.

Stage Two: method of lowest risk

Prior to the work commencing the pre-contract period should be used to 'tune' the programmes and further investigate the methods of working with regard to safety. A policy of achieving a 'method of lowest risk' should be adopted when devising the detailed working methods. Each activity and operation should be considered within its own unique circumstances. Although many activities can be seen as repeats of previous ones from job to job, their position, place in the sequence, environment, and the presence of other hazards, can make them very different operations.

Stage Three: defining responsibility

In the written working method statement the people responsible for the separate activities are to be named together with the individual who is to take overall responsibility. The named people should be informed and told exactly what their roles are in the process.

Stage Four: implementation

Initially, effective communication should be ensured: all people working on the activity must know exactly the stated methods. No deviation should be allowed from the statement. Where to follow the statement would be impractical/impossible, work should stop and immediate reference be made back to the author. Only after a written variation has been issued to all people can work recommence.

Stage Five: development

No system can ever be said to be perfect and therefore those people involved in implementing the working method should be encouraged to offer ideas for improvements or to criticise those aspects they deem to be unacceptable. This information should be passed back to the originator and site and office management to help future methods. Unfortunately, feedback on safety matters only tends to surface after an accident. We need to also note the near misses and seek the active support of those people carrying out the hazardous operations.

To sum up, safe working must be seen as a fundamental ingredient of any production process. The aids used in the process must be designed primarily to provide safe working at all times.

6. Plan and integrate resources and elements to produce process

The basic working methods and sequence have been devised, the resources ascertained and the safe working aspects considered: these are now brought together and matched to availability. When drawing together the parts of the construction process it is likely that three situations could develop: one, that a solution evolves which is derived from common practice using basic skills, plant and methods; two, that basic skills and plant are used in an unusual manner; three, that a completely new skill or piece of equipment has to be developed in order to meet the demands of the building's technology.

By analysing the building afresh each time more confidence in producing the optimum production process can be achieved. It might be that major adaptations bring an unusual method into line with common practice, thereby reducing costs attributable to on-site learning and the possible extravagant use of resources. The reverse could also occur: a basic method will not produce the desired results and an investment will have to be made in a new method or piece of equipment. A major factor in the adoption of this solution is cost – will the results justify the cost? Can the method, equipment be used on other jobs? Another factor which can override cost is that of safety. It may be necessary to adopt an expensive method because of the potential hazards.

The production process is in essence the integration of resources, aids and elements, and the plan should take note of their ability to meet the demands of the building when required.

7. Monitor process

The monitoring process should continuously update the availability of resources and be aware of new developments in statutory regulations, working methods, and plant and machinery. Even where a working method has been agreed it may be advantageous to alter it to make use of a new piece of plant or equipment. There are continual developments in systems and the plant to execute them: the building technologist should be investigating and evaluating them in relation to ongoing activities. The monitoring process should be carried out on site and involve all those people doing the work. The long- and short-term availability of resources should be checked constantly: the rate of working against the programme should be assessed weekly, and any deviations should then be related to the planned sequencing of resources. Where new methods or plant are to be introduced this should be with the full knowledge and participation of all the people involved: adequate training to be given where necessary. A system seen to be productive on paper can cause losses in time, materials, energy and costs if not properly introduced and implemented. If the new system is found unsatisfactory it may be better to return to the old, even if a loss of face is incurred.

8. Record data and feedback into future processes

Irrespective of whether the production process is successful or not, whatever was learnt must not be disregarded or lost. Unfortunately, owing to the fact that, in most cases, on completion of the work the design and production teams split, retention of data and its dissemination to all is not easy. Within each team a further split occurs. Therefore, the task of collecting and recording data is extremely difficult. In the case of an unsuccessful project this difficulty is compounded by those involved not wishing to impart information which could show up their contribution to the débâcle. Any information recorded should be saved within a system for the planning of future projects. It should be easily retrievable and be presented in such a way that it makes decision-making a logical process. For example, if a system of formwork has been developed, with a method of working and safe practice, this should be recorded so that the technical information is clear and concise and fully described, with diagrams etc. – the source of the equipment and materials is noted; the method of erecting and dismantling described; particular safety hazards mentioned; any cost advantages given; any drawbacks or disadvantages highlighted.

The principles listed in Fig. 2.3 should be applied even where the building form is well known: nothing should be taken for granted. If the result of the analysis is the same as before it will still have been a worthwhile exercise, for it will confirm that the present methods are still optimal. Some of the most common building forms will now be considered in the light of the above eight principles.

Frames

There are three types of frames under consideration, namely beam and column, steel and portal.

Beam and column

Timber is used in some structures, predominantly in framed panels as well as being laminated in portal frames.

Assume the building is a fifteen-storey frame. Most beams and columns will arrive on site piecemeal and be ready prepared for lifting into position. Therefore one necessary aid to production will be a lifting device. Access to and around the frame members will need to be considered. Primarily there is a need to recognise structural form (Principle 1).

The secondary elements will then be investigated. What are the floors? What is the cladding? How do the services fit in? Is the roof posing any particular problems? (Principle 2).

The primary and secondary elements will need to be considered together. Do the floors need to be formed simultaneously with the frame or immediately after? Can they be left till after the frame is complete? Do the floors or other secondary elements contribute to the structural stability of the building? How much of the frame can be erected at any one time? (Principle 3)

It is now known that lifting appliances are required for the frame's members. What are their sizes and weights, and to what height are they to be lifted? Is there access around the building? What plant and equipment is required for the floors and other secondary elements? Can one piece of plant be used to aid production of more than one element? Will the required speed of erection affect the number and type of plant? What skills will be required? Generally, specialists fix the frame. Do they provide their own equipment? (Principle 4)

Even though specialists erect the frame it is the builder's responsibility to ensure they work safely. The specialists should supply a written working method statement to the builder and carry out the work as stated. The builder should monitor the safety aspects to make sure, for example, that the steel erectors are wearing safety harnesses and that they are attached to a safety ring. The builder should prepare a working method statement for his operatives when working adjacent to the steel erectors. Will work be allowed below the areas of frame being erected? How will this affect the installation of floors or cladding? (Principle 5)

Now that a clearer picture of the building's technology is known, from the answers to questions posed above (amongst many others) the integration of materials, technology and aids to production into a production process can begin. The aids should create a flow and rhythm which can only be achieved by careful planning and execution on site. If the floors are precast, can a crane lift these as well as the steel members? If in situ, where does the concrete come from and how is it placed? What activity follows on? What areas/sections can be released (and when) for the following operations? What plant and equipment can be moved forward to ensure a smooth flow and optimum usage? What can be done to improve methods? (Principle 6)

When work is in progress the operations need to be constantly monitored in their own right and also in relation to those following. All resources will have to be organised and brought into play at the appropriate time and place. The monitoring will be the responsibility of site management, together with those who drew the initial plans. Where improvements can be made they should be encouraged, but not at the expense of quality, safety or cost. Modifications to the working methods should be agreed with the initial planners and/or senior management. For example, it may be that a section of the steel frame is behind schedule owing to a break in deliveries: work is then speeded up on another section. A revised working sequence will need to be prepared – can follow-up work still proceed in the delayed section? Will it mean the reorganisation of plant and equipment? Will the structural stability of the frame be affected? How will this affect the deliveries of materials required for the secondary elements? (Principle 7)

Finally, the working methods and technological solutions should be recorded. Photographs, diagrams, notes and trade literature, together with a critique of the system, should be filed where they can be retrieved for future planning. How did the delay in erecting a section of the steel occur? How did it affect the production process? What are the lessons for the future? (Principle 8)

Steel frame

To revert to the steel frame, the following is a summary of the production process, bearing in mind that each of the eight principles have been considered in the evaluation. The frame is fixed together on site, being primarily an assembly process. It commences from ground level and proceeds upwards floor by floor, although some sections can go in advance of others. The basic aids to production are lifting appliances able to reach to the required heights. Space will be required to position the lifting appliances and provide short-term storage areas. The same plant could be used for the placing of precast floors. Safety barriers/access may be required if the cladding does not follow immediately upon the frame and floors. If in situ concrete floors are used a different site technology will be used, involving formwork and the handling of wet concrete. Is the formwork permanent or temporary? How much reinforcement needs to be stored and placed? If the concrete is lifted can it be carried out by the same lifting appliance for the frame? Will it be better to pump the concrete? What is the source of the concrete, on-site or off-site?

What methods are used to protect the frame from the effects of fire? Will this have to be done during the initial stages of the project or can it be left to the finishing stages? The means of protection can have a profound effect on the site processes. For example, if concrete is used, formwork will be required and pours will be in small quantities. Is any other concrete required on the job? How does this affect the final plumbing and securing of the frame joints? Do they have to be kept free until all the frame is complete? If the fire protection is of a spray or board type, it will need to be applied after the building is watertight but before many of the finishing stages.

If the external cladding is lightweight panels and glazing, how is this fixed – internally or externally? Are access platforms required? Does a scaffold need to be erected? Can cradles be used? If so can they be the ones that remain on the building for maintenance? How is the cladding lifted into position? Will the lifting appliance for the frame be able to cope?

What services are specified? How can they be integrated into the production process? Are they part of the structural system, or seen as part of the secondary elements? Are the lift shafts an integral part of the structure? What influence do the stairs have on the building's structural integrity?

As the production process for a theoretical building cannot be defined the summary ends in a series of questions. It is by asking the right questions that the production process will evolve. Accordingly, some general points to provide guidance for production will now be given for other types of framed structures.

Portal frame

Portal frames can be in steel, precast concrete or timber, and up to two storeys in height. They can be placed using a mobile crane. Large spans may require two cranes working either side of the building, or an overhead gantry moving on rails, for long buildings. The cladding and roof are fixed independently of the frame. Depending on the type of cladding will be the type of access: scaffold or scaffold towers, either internal or external. The prime aid to production is mobile lifting appliances with relatively long reaches, followed by access for fixing cladding and roof.

Precast concrete frames are larger than steel frames. Similar problems arise, but with the added complication that joints between members have to be grouted after fixing. Formwork may be required to hold the grout in place whilst curing takes place. It is common for floors and roof to be also in precast units. The operation becomes one of assembling and grouting the units together. Minimal external access is required, depending on external cladding. As the units are concrete no additional fire protection is needed, giving them an advantage over steel frames.

Timber frames are usually confined to two- or a minimum of three-storey housing. The frames arrive on site in the form of storey-height wall panels, up to 3 metres in length, including openings. Floors can also be supplied in panels, with boarding in place. Trussed rafters can be joined on the ground, together with sarking felt and battens, and lifted on to the structure. This method will require a mobile crane. It is possible to erect a two-storey house by hand only, if the panels are manageable by two people. A scaffold will be required to give a platform for the handling of the first floor panels and for the placing of the individual trussed rafters. External cladding has to be applied to the panels to give them adequate weatherproofing, insulation and fire protection. If this is in brick an access and storage scaffold will be required. If using timber boarding it may be possible to work from mobile scaffold towers, depending on ground conditions.

Load-bearing walls

Generally, the on-site operations required to form load-bearing walls are more complicated than for the joining of framed structures. There are more separate tasks involved in erecting in situ walls. Precast concrete walls have been used, but are now not so common in the UK owing to poor quality in their manufacture and fixing. The usual materials are blocks, bricks and concrete. Sixteen-storey structures in bricks and blocks can be built. The aids to production are: materials storage on the ground and in close proximity to the point of use; supply and mixing of mortar; movement of mortar and bricks etc.; access to one or both sides of the walls; supply and placement of lintels over openings; constant supply of water.

Concrete walls require further aids, over and above those employed for brick walls: materials storage in separate bins, and adequate protection; supply and mixing of the concrete under quality-controlled conditions; movement, both horizontally and vertically, for the concrete; access to the point of pour; constant supply of water; steel reinforcement; formwork to hold the concrete in place during pouring and curing. The big element in this process is the manufacture and erection of the formwork. The speed of erection will dictate the speed of concrete placement, as placing the concrete is a relatively fast operation. It can take weeks to make and place formwork and the pour be completed within a day. When using load-bearing walls the floors must be placed on them before commencing the next storey, since the floors need to be directly bearing on the walls to avoid eccentric loading. Support nibs or corbels can be used but this will incur extra formwork. Eccentric loading of brick and block walls is to be avoided. Another complication in the production of an in situ wall in brick or concrete is that it cannot be loaded until some time after being built: prefabricated walls and frames can. The in situ materials need to cure and gain strength over a period of time. If the wall is loaded soon after erection then the imposed loads should be temporarily propped to take their weight. This will require further production aids such as adjustable props.

Core structures

A central core is usually constructed in advance of the rest of the building. It can contain the lift shafts, stairs and toilet facilities for the new building. The core will take all the loads of the building and transport them to its foundations, although the outer structure may derive some support from its own columns/walls bearing on foundations. A need to erect tall structures for radio/TV transmission produced the technological solution of slipforming. The formwork to create the shape of the core structure is constructed at ground level, be it circular, oval or rectangular in shape. Using a system of jacks the formwork, in one whole piece, is slowly lifted as the steel reinforcement and concrete is placed. As the concrete cures it can give support to the slipform equipment and enable it to climb on the structure. In theory the process can be continuous and go to extreme heights. The governing factors are speed of fixing of the reinforcement, time for concrete to acquire enough strength to support the slipform equipment, and ability to lift the concrete to the height of the core. The slipform equipment can carry with it cranes, passenger hoists, concrete pumps, access platforms and safety items such as nets and fans, and can give limited storage. The production mode is the slipform equipment and this has to be designed and built to meet the particular requirements of the core. The equipment used on the Nat West Tower in the City of London won an award as a steel structure in its own right, even though it was dismantled and scrapped after use.

Using slipform on core structures confers two main benefits: one, that the work can be continuous and therefore fast; two, that all equipment is prefabricated and lifted in one piece. For this method to be economic the tower needs to be tall and/or to be built in an extremely short time.

Other structures

Tension structures and other uncommon types remain to be mentioned. Tension structures can be formed in two ways: either where the cladding/roofing is brought into tension over supporting columns like a tent and secured to the ground, or where columns are supported and tensioned by ‘guy ropes’. The columns support the roof, and if necessary the cladding. Precise engineering skills are required for setting out and ensuring the correct tensions are orientated towards giving temporary support to the structure whilst applying the tensions.

MATERIALS PRODUCTION

Many materials arrive on site in a natural or roughly shaped form and are worked upon before or during placement. In the past the great majority of items and components were formed in place or prefabricated on site. During the 1960s, on a large housing contract (1100 dwellings), the following operations were carried out: all mortar was mixed for brickwork and blockwork, using sand, lime and cement; all the timber window and door frames were made from planned dimensioned sections in a temporary workshop on site; concrete lintels were prefabricated on site;

approximately half the roofs were formed using rafters, joists and purlins etc. cut in place. Nowadays the incidence of these methods of working will not be as high, but many types of building work, such as refurbishment and extensions, will involve a high amount of cutting, mixing or shaping on site.

Three materials – mortar, concrete and timber – will be considered to ascertain how they determine their aids to production.

Mortar

When producing mortar there is a need to satisfy a range of factors, from ensuring the correct mix of materials to transporting it to the point of use. The production process has first to establish the fitness for purpose of the materials in their unmixed state and then to carry that fitness to the point of laying; plant will need to be used and a system of working devised. The process is described below, stage by stage.

Storage. The materials need to be stored in separate bins on a firm, even base. Cement should be in a properly constructed store shed (or silo for bulk).

Mixing. The mixer must be adequate for the purpose. It must be able to mix to the right quality and quantity. High quality mortars can be produced from pan mills, large quantities from reverse drum mixers. If the mortar is moved by vehicle the mixer will need to be placed off the ground so that the discharge can drop into the vehicle.

Transport. Ideally the mixer should be as close to the point of use as possible. The transporting vehicle should be the optimum size for the rate of usage: too big and there will be waste, too small and demand will not be met. Where vertical movement is required a fork lift can be used (up to three storeys height); it can also provide horizontal movement. Purpose-made transit bins should be used for extensive travel. Protective covers should be provided in extreme weather conditions.

Placing. The mortar needs to be removed from the transportation vehicle or bin to the work 'spots'. A dump truck can deposit the mortar onto a large boarded area from where it can be moved by hod. Bins can be lifted directly to the point of use.

Curing. Some protective measures may be required to control the curing of the mortar in the wall. Waterproof sheeting should be used during heavy rainfall, damp sacking under hot sun, and insulated quilts in cold weather.

A theoretical building development will now be described and the options for production of the mortar discussed. The site is a housing estate totalling 240 dwellings in two- and three-storey brick and block load-bearing wall construction. The work is to be continuous and phased handovers are required; speed is essential. It is expected that at least eight gangs of bricklayers will be working at peak periods. The first option to be decided is whether to produce the mortar on site or buy it in ready mixed, either ready for laying or in a form where cement and/or water needs to be added. If it is decided to produce on site the following questions will need to be answered. Where will the materials be stored and mixers placed? What type of mixer will be required? How many mixers? (Geographic distribution of the dwellings, number of bricklaying gangs, speed of work will need to be brought into the reckoning.) How is the mortar to be transported? Can the scaffold take bin loads? A central mixing plant, comprising two pan mixers, material storage areas, cement shed and water supply could be set up. From here dump trucks or fork lift trucks could distribute the mortar around the site to all bricklaying gangs. A problem with this is that it is difficult to meet demand at peak periods, such as morning starts and after meal breaks. An early start for the mortar production crew would be required to ensure that all gangs had mortar at the start of the work period. An advantage with this set-up is that greater control can be exercised over quality and lesser costs can be distributed over all the gangs. For eight gangs a five-man crew (two on the mixers, two driving the vehicles and one in the cement shed) can service them. If each gang has its own mixer then eight people will be required to produce the mortar. But there are advantages in using separate mixers: transportation distances can be minimal, and each gang can have absolute control over the quality of mortar. Disadvantages would be that extra mixers will have to be paid for and separate water supplies set up for each, with separate storage areas and cement sheds.

The second option is to use premixed mortar from an off-site supplier. Each load would have to be up to 7 cubic metres to be economic. It can be distributed to the gangs by using small bins, this being done by site transport. If a half premixed supply is used (where water and/or cement is added on site) then mixers would be required. Again, there is a need to consider whether to have one for each gang or a central set-up. Perhaps a compromise solution would be effective, e.g. two or more gangs using one mixer. This might be possible, depending on the size of the gangs and the quantity of mortar they require.

An optimum solution will be reached by balancing the following factors:

- quantity of materials required at peak times
- production capacities of mixers (types available)
- geographic layout of the site (sequence of building, space for mixers, types of terrain etc.)

- specification of mortar and constraints in its use
- distances between mixers and point of use
- methods of transporting the mortar, horizontally or vertically
- preferences of the bricklaying gangs
- costs (premix as compared to site mixed, many mixers as compared to a central mixing plant)

Concrete

The factors involved in the production of concrete are similar to those for mortar, so may be taken as a starting-point.

A major choice to be made is between on-site and off-site production. The most common reasons for not using site-mixed concrete are: lack of room for storage and mixing plant; the quantities of concrete are small or are required in small infrequent batches; the economics of setting up the plant are not acceptable; concrete-mixing skills are not available. Even where site-based batching plant is used it is not uncommon for this to be augmented by ready-mixed supplies. The rate of production from an average-sized site mixer is not high and on large pours a constant high rate can be achieved by well-controlled deliveries of ready-mixed concrete. The amount of concrete required, its quality and demand rate, is determined by the building's technology. For example, concrete used as fire protection to a steel frame will be required in small quantities over a long period of time. Although the total quantity may seem high, making site-mixing appear an economic proposition, it may not be appropriate as the plant will be standing idle for long periods. It would be convenient to match the concrete pours to the size of ready-mixed lorries, always assuming that they are within travelling distance of the site.

Another factor which could influence the decision regarding on- or off-site mixing is that of quality control. The builder may want to produce his own concrete to ensure that quality is correct and control procedures are satisfactory. Also, any planned variations in the mix can be more easily achieved. Furthermore, greater flexibility in the production of the concrete can be realised – at a moment's notice at any time of day.

The movement of concrete on site can be organised in a number of ways: via a pump and pipes; or a skip handled by crane; or a dump truck or revolving drum on a lorry. Modern pumps are capable of transporting concrete over considerable horizontal and vertical distances at a continuous high rate. Concrete carried by skip (normal capacity 0.5 to $1m^3$) is dependent on filling the skip, lifting it to the point of use and returning it to the discharge source – a slow process. A dump truck or lorry can carry large quantities over horizontal distances and is ideal for placing concrete in foundations and ground works. By using chutes the concrete can be directed to the point of use.

The production process of forming concrete structures can be complicated by the need to place large amounts of reinforcement. The reinforcement may be so extensive that a special mix has to be designed to flow around the bars and special vibratory equipment used to consolidate the concrete. In that case, further aids to production will be required.

Curing the concrete may also raise some problems which will have to be solved by additional aids. In winter some form of insulation may be needed for materials stored on site and for formwork. In hot weather the concrete should be cured under a damp membrane.

Overall in the production of concrete, the following need to be determined:

- quantity of concrete and rate of requirement
- production capacity of mixers or ready mixed deliveries
- positions of concrete placement, horizontal, vertical constraints
- specification of concrete mix
- method of transporting concrete
- problems in reinforcement, formwork or weather
- space available on site
- skills required
- costs

Timber

Many types of building work require timber to be delivered to site in a semi-prepared state. If used as carcassing the timber will be sawn (if structural it will be stress-graded) and joinery timber will be planed all round. Timber is generally cut to precise length on site, although it will be ordered in lengths which are close to the finished requirements.

In accepting that timber has to be cut to length on site, how best can this be achieved? The traditional method is to use a handsaw: a fine-tooth saw for joinery timber and a coarse-tooth blade for carcassing timber. If there is a large amount of cutting, especially on large-section carcassing timber, then it is practical to use a power saw. This should be a circular saw, fully equipped with safety devices and only operated by a trained person. It should be positioned in a central place (mobile units are preferable as they can easily be repositioned). Timber storage should be off the ground in well-ventilated shelters adjacent to the saw.

Increasingly, use is being made of small, powered hand saws for joinery timber, both circular and jig saws. Whereas a large circular saw will be provided by the builder, small power saws will usually belong to the operatives. They make the work less tiring and speed up the process. Power will be required, as most run on high voltage electricity (preferably 110 volts). Therefore, power sockets will need to be provided, together with distribution cables.

The use of hand tools allows the carpenter to control the pace and position of the work. With large power saws the carpenter has to work near them, while with small power saws he is restricted to the run of the cable. When using power tools some thought and planning has to be given to ensuring that they are fitted into the method of working. Whilst they give, literally, more power to the operative, the technologist has to plan for them and control their use. The management responsibility for providing the means to achieve the desired ends, following the trend towards mechanisation, is discussed in the following section.

MECHANISATION OF SITE PROCESSES

The great advance in the provision of aids to production has been the replacement of physical labour with mechanical devices. In reviewing these aids the substructure will be considered first, and then the superstructure.

Substructure

The work involved in digging out the ground and moving the spoil on and off the site has been a natural arena for major advances in mechanical aids. Spoil is heavy and cumbersome and any means to reduce human labour in its handling is of great benefit. The range of machinery now available is extensive and far exceeds the choices for other types of construction activity. The reasons for this include:

- relatively high cost of labour in industrialised countries
- the development of machines for particular excavation tasks
- the quest for greater speed in construction
- the quest for greater productivity
- the complexity and dangers of foundation and basement excavation.

With the introduction of versatile and efficient digging machines, whether backactors, front shovels or grabs, the excavation of deep basements and complex structures below ground level has been made possible. For example, the hydraulically operated crawler-mounted grab has enabled diaphragm walls to be excavated: a deep and narrow trench can be accurately excavated under a bentonite slurry.

How does the use of plant affect production? A number of aspects are involved: speed; technological constraints of the excavation; access and movement around the site; disposal (or otherwise) of excavated material; skills available for plant operation.

Speed. The excavation and removal of soil can be carried out extremely quickly by machine as compared to hand labour. Where quantities of excavation are small the costs of bringing a machine on to site may outweigh the possible gain in time. Times for excavation need to be integrated with other activities. What dictates the speed of substructure work?

Technological constraints. Many excavations need to create multi-floor levels below ground level. Restricted sites will also pose problems for machinery, as will old or existing structures in the ground. When excavating deep or multi-level basements or foundations it has to be decided whether the machines follow the dig down or remain at one level. How is the working area protected? What other constraints might there be on the use of machinery? Does this restrict the type of plant that can be used? How does the nature of the soil affect the choice of machine?

Access and movement. Is there enough room for the effective placement and possible movement of the plant? Can the plant get to the site? Do support platforms and piers need to be constructed? What movement corridors need to be designated? If there are a number of machines on site will they hinder each other?

Disposal of excavated material. The type of machine employed may be determined by the method of spoil disposal and movement. For example, if spoil is to be moved over a long distance, either a large bucket should be used or special earth-moving equipment. Is it better to load directly into removal vehicles or create spoil tips for future removal? Can wheeled vehicles move over the ground in all conditions?

Skills required. There is a need not only for well-trained machine operators but for suitably trained people to direct and work with the machines. A piece of machinery is totally reliant upon the people who operate it. The requirements are:

- to know the machine's basic working elements and principles
- to be able to operate the mechanical functions of the machine
- to know the limitations of the machine
- to operate the machine safely under all conditions
- to be accurate and to work at an efficient speed
- to work effectively with other machines or operations
- to carry out day to day services and checks to the machine
- to forecast the output and work capacity of the machine, together with the ancillary plant
- to select the appropriate machine for the defined task.

Some of the above skills must be possessed by the operator, whilst others are the province of the technologist. It is the effective combination of all that will achieve efficient, safe working on site.

Saturation point in the development or range of machines has not been reached. A recent development has been the manufacture of small diggers and material movers capable of working in extremely confined spaces. They are self-contained and mobile, with some able to pass through the width of an ordinary door opening. Piling plant is constantly being improved, giving quieter operations at greater depths.

Superstructure

The aids to the construction of the superstructure – its structure, fabric and finishes, and services – are, generally, not as direct in their input as those for substructure work. Where substructure aids are seen as prime movers, those used in work above ground make their contribution in less dramatic but no less important ways. A crane (tower or mobile) is useful for lifting materials and components. A compressor can provide compressed air for power tools which enable a job to be done easily and relatively quickly. Fork lift trucks can move items around a site and lift them to storage and working platforms. A hoist will lift materials from the ground to any floor level.

An example of a direct mechanical aid in superstructure work is in the lift slab method of construction. This is based on the principle that it is easier to cast floor slabs for all levels on the ground; they are then jacked up to their resting places. The columns of the building are constructed first. The floor slabs are cast one upon another with separating membranes within the building itself and around the columns, to the size and required shape. They are then jacked up using the columns as supports and guides. The uppermost floor cast becomes the top floor of the building. This method is totally reliant on mechanical aids. Likewise, slipform techniques are dependent on mechanical lifting devices and control apparatus.

Simple production aids for the superstructure have been used ever since buildings above two storeys began to be erected. The block and tackle is one example, which is still in use even on sophisticated types of building work. Wheelbarrows are used for moving materials around. Progression has taken the form of replacing human with mechanical effort, in the use of engines with the block and tackle. Further advances have taken place in the efficiency of the engines and in the complexity of the lifting arms so that very large weights can be carried.

Another example is the forming of a wall in panels of precast bricks, allowing the physical labour of laying the bricks in situ to be replaced by mechanical lifting apparatus. Along with the trend towards fabricating many components off site is a corresponding move towards the mechanisation of site processes. Components will be delivered to site in a near-finished state and will require placing and fixing. The size, shape and degree of protection required will produce site handling problems: these will best be solved by mechanical means.

EFFECTS OF MECHANISATION ON PRODUCTION

Whilst it can be said that mechanical aids make the work easier they do also add to the complexity of construction operations. The effects are listed below.

Speed

Most operations can be carried out much quicker than by hand. As further improvements are made in machinery it is likely that they will speed up the processes even further. An example is the placing of concrete. This was generally carried out by skip and crane before the advent of pumps. The early pumps were limited in their reach and the distance they could send the concrete. Modern pumps can send concrete to great heights and at a fast rate. It is likely that their performance will improve still further. Speed can cause difficulties which require further aids to production in their solution. Pumped concrete, for example, can be placed in large quantities at a rate so rapid that the initial set is taking place when upwards of $100m^3$ has been poured. Therefore, support formwork has to be designed to take dynamic loadings.

When planning and sequencing construction activities the speed of task achievement should be ascertained in relation to the ue of machinery. It may be more economic to select a smaller, slower machine that keeps pace with the general level of activities.

Energy

All plant and machinery require some form of energy input, which is commonly provided by electricity (mains or generated on site) or petroleum fuels. Mobile plant tends to use petroleum fuels whilst static plant uses electricity. Diesel engines are more powerful than electrical motors, size for size, making them ideal for moving spoil etc. Mains electricity requires a cable supply, which can restrict movement.

Adequate storage and distribution facilities need to be planned and installed for the supply of the energy. Getting the energy to the plant must be considered alongside its placement and movement.

Atmospheric pollution is becoming an increasingly important factor to be considered. The use of hydrocarbon fuels creates waste which enters the atmosphere, and public concern is growing about the effects this may have on health. In confined site areas, for example, diesel engines may create a need for mechanical ventilation.

Complexity

With the increase in the number and types of machinery and plant available one could be spoilt for choice. The ideal way to select plant is to consider the particular operation to be carried out and then to analyse it. First, a performance specification is drawn up to set the parameters for the task. This will give the criteria for meeting the demands; against these can be matched the means for achieving them. It is possible that no one machine will be able to meet all the criteria satisfactorily and that a number will come close to the optimum. In this case other factors need to be considered, such as the versatility of the machine, the fuel required, maintenance levels, running costs, the need for specialist operators, capital costs/hire costs, availability. Unfortunately much plant is selected not on the principle of matching task to machine but according to what is available at the time and what can carry out a multitude of tasks. This is understandable, but the machine can then be put to use on tasks not suitable for it, with consequent extra wear and tear and perhaps unsafe working procedures. One instance is the use of a tractor with fork lift tines to carry plasterboard. If the boards are placed directly on the tines they will bend and whip as the tractor moves over rough ground. Not only is damage caused to the boards: they could easily fall off and perhaps hurt someone.

When this machinery and other aids are on site one needs to be able to control them effectively. Machines are complex in themselves and when geared to the achievement of high levels of productivity the complexity of their operation increases. When a machine is substituted for a man that machine has to be programmed into the site activities: it cannot integrate itself.

6. Temporary Works: Falsework

PRINCIPLES

Most falsework is prepared to hold the forms for wet concrete, although it may be used in some instances as temporary support for existing structures or prefabricated units. Formwork is the actual mould for holding the wet concrete. For the purposes of this discussion the two overlap and principles and practice will be applied to both. For example, a waffle floor formed with glassfibre moulds resting on ply sheeting must be supported by falsework, and the method of forming the floor cannot be divorced from its support when designing the whole. There are four major principles that need to be considered when designing and providing falsework.

1. Safety

Following a series of accidents in the UK and other countries in the early 1970s, where falsework collapses led to serious injury and loss of life, a committee was instituted to consider the reasons for such collapses and to put forward recommendations for their avoidance in future (I.L. Bragg, *Health and Safety Executive. Final report of advisory committee on falsework*, June 1975, HMSO). Its report made clear that in the design and provision of falsework safety aspects must become central. Adequate and proper design procedures must be followed, based on sound structural theory and analysis, and site erection processes must be carried out strictly to the design parameters.

2. Access

Access must be provided for the operatives to get to and place the concrete (or prefabricated units). Consideration of this should not be confined to the actual place of work but also to the area surrounding it. Can the operatives easily approach the place of work? Can the necessary plant and machinery be brought to the point of use? Are the access platforms strong enough to carry the loads?

3. Support

All design for falsework should be carried out by a qualified person using sound structural theory and analysis and calculations. The materials used must be of the highest quality.

4. Aid to production

As far as the builder is concerned falsework is a direct aid to production, one that can account for up to 60% of the total cost of forming an element. In addition to the cost factor the builder must bear in mind these factors: utility (can the materials be used again?); convenience (are the materials readily available? how do the design and materials fit in with the other aids to production?); speed (can the falsework be easily and speedily erected and dismantled?)

Application of these four principles can be confused by the split in responsibilities for design and erection, as was found by the Bragg inquiry. Some problems identified by the committee were: access openings for contractor's plant were introduced after the initial design was completed; preliminary design was undertaken at the tender stage when only preliminary documents were available, but once the contract was awarded there was tremendous pressure to produce the final designs at high speed; sometimes the need for changes was not communicated back to the designer and mofidications were made on site which seriously weakened the structure. Many of the communication and procedural difficulties encountered in the UK were associated with the way that the industry is organised: most of the tasks are carried out in small groups, on an ad hoc basis, on a great variety of construction sites. The report recommended procedures that ought to be adopted to deal with falsework, namely: checking and cross-checking of the design; inspection of the falsework; and, above all, the maintenance of good communications between all parties. It also recommended that a Temporary Works Coordinator (with defined duties) be nominated for all construction involving falsework. Fig. 2.4 shows an organisational structure and task responsibilities geared to achieving a strategy within which proper design can take place. Everything must flow from and through the coordinator, who may also be the designer.

In choosing contractors the client should consider their record with regard to safety and training. The report says that a client should be free to prevent a contractor from tendering if he is believed to have too little regard to safety. Tenderers should be able to support their claims regarding safety by reference to low failure rates, and a satisfactory training programme.

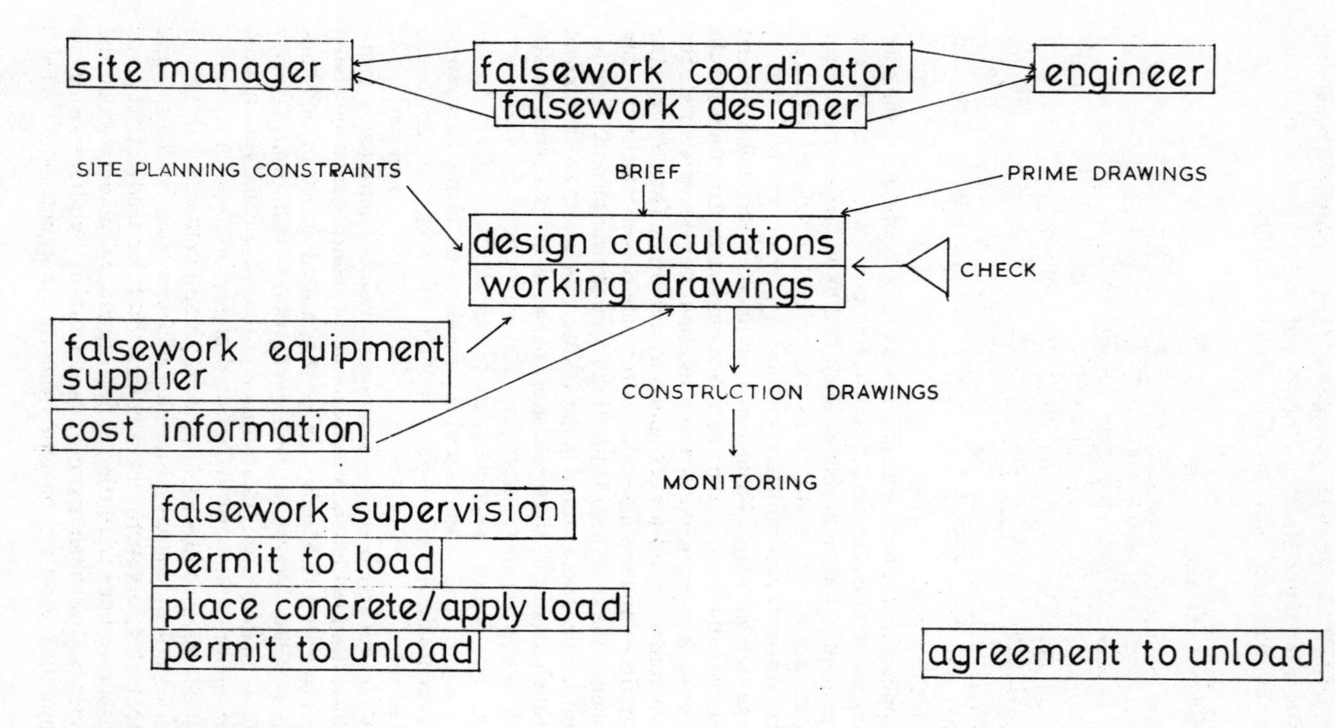

2.4 *Co-ordination of falsework*

DESIGN

The design of falsework should take into consideration the following factors.

Estimation of loads

There are problems in calculating the loads on formwork: the concrete density may be higher than envisaged; adequate distribution of the load may be temporarily prevented by reinforcement producing concentrated loads where such were not expected; the process of building the permanent structure may prevent an even load distribution; critical loads on falsework may occur while the concrete is being placed. The designer should prepare a loading programme when large quantities of concrete at high rates of pour are being placed.

Concrete can produce lateral forces on side shutters which can cause sudden movement. A vertical shutter resisting horizontal pressure tends to rise despite the fact that there is no apparent vertical force to cause this. In consequence the shutters should be effectively anchored down.

The effect of wind on falsework is greater before and during pouring than after the concrete has been placed. Attention should be paid to the geographical location of the site and the cross-section of falsework plus framework. The denser the supporting falsework, the less streamlined the component; the greater the area of formwork, the higher the resistance.

Some allowances should be made for impact, such as that caused by shock loading or during the application of the load, and for the need to provide access openings.

Lateral instability

Since lateral instability is a frequent cause of collapse an increase in the calculated loadings is required. In practice this means that the falsework will have to be properly braced, guyed or tied back so that it is stable against lateral and longitudinal forces. To give lateral stability, sufficient bracing in plan, transversely and longitudinally, must be designed. Bracing between tower units, foot ties and head lacing is especially important.

The designer should bear in mind that the falsework, and perhaps the formwork, will not be erected perfectly vertical and without eccentricities: the design should cater for normal site tolerances, which should be clearly indicated on the drawings. If proprietary units are proposed then the designer should have a proper understanding of their performance characteristics. Where these are integrated with other systems their compatibility should be considered. Again, it is important to be aware of the need for bracing of proprietary units, as trade literature may not show it to be a requirement. Partially erected systems may be vulnerable to instability, and a proprietary system should be designed in such a way that as far as possible all dangers of incorrect assembly are eliminated.

The factor of safety used in the elastic design of structures should be 2. Where the design and construction method are novel and the working conditions poor, this factor of safety should be nearer to 3.

Fig. 2.5 is a flow chart for checking the design of falsework. Although this should be drawn up by the design initiator, the calculations and execution of the proposals should be checked on site.

Site provisions

The falsework design needs to be implemented safely and efficiently on site. There are three factors pertinent to site operations which complement the four principles stated earlier (pages 89–90).

MATERIALS

The choice of materials and their integration into the formwork structure is a prime responsibility of site management. The design is influenced by the materials available. The builder may have certain items in stock and have arrangements with particular manufacturers/hirers of falsework systems. The choice of materials may also be influenced by the skills and experience of the falsework erectors. Normally, these variations are acceptable but where the use of readily available materials is inappropriate those demanded in the design must be used.

There are standards published, in final or draft form, from a number of countries. The principles they propound should be included in UK practice. For example, stress-graded timbers should be used where loads are to be applied upon them.

It should go without saying that all materials should be of the highest quality. Where fittings, bolts etc. are used they should be free from corrosion and/or damage.

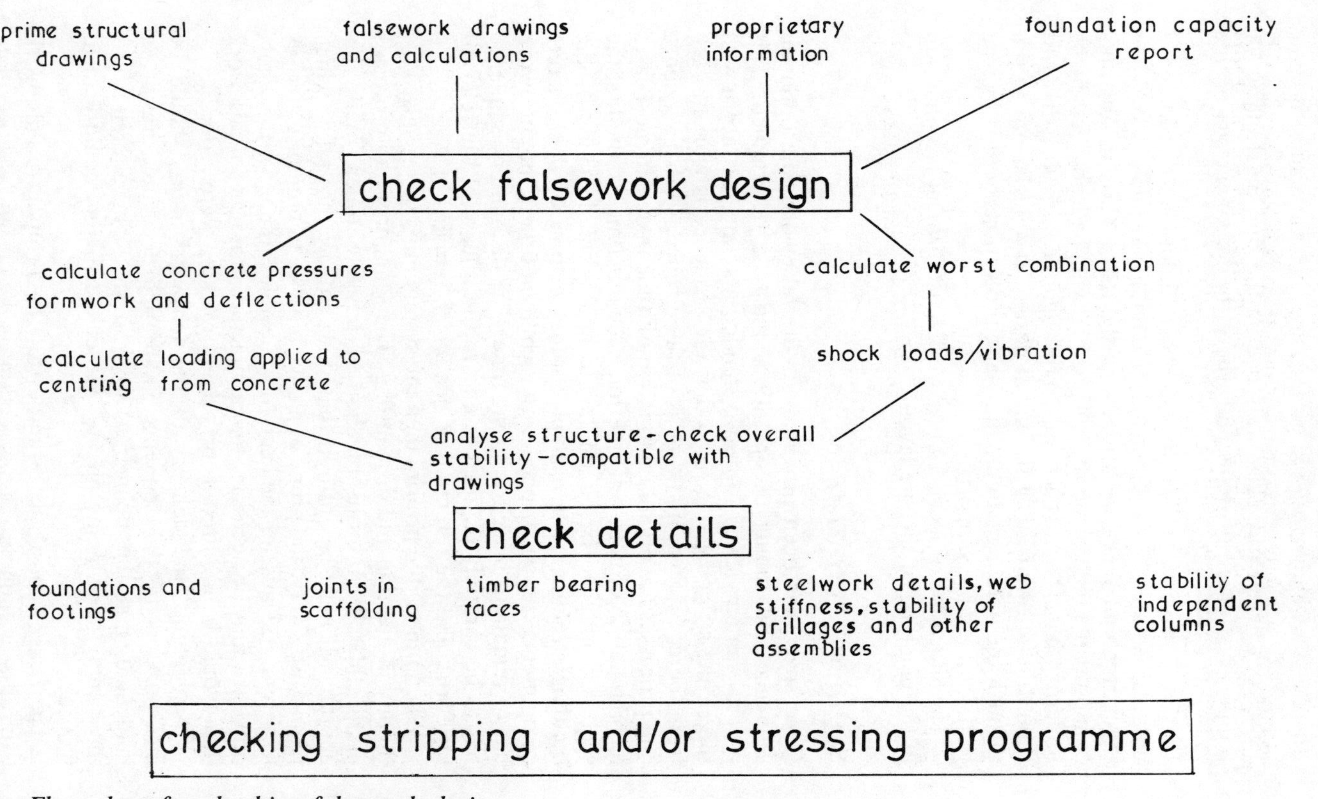

2.5 *Flow chart for checking falsework design*

SPACE

Falsework necessarily occupies space on the site, particularly for temporary storage of items and where the falsework sits on the floor. This precludes the use of that space for other activities. Indeed, the space under the falsework should be considered a no-go area unless work is being carried out to the falsework. When the partly built Westgate Bridge over the River Yarra in Australia collapsed it killed a number of people who had no need to be underneath it. If access is required through the falsework then this should be allowed for in the initial design. If vehicles are to run through the opening, strenthening and protection should be given to the vulnerable supports.

The allocation of space for falsework must form an essential step in the planning of the site layout.

SEQUENCE AND TIME

Once space has been provided for the falsework, the optimum sequence and timing of operations must be determined. This process can be viewed from two standpoints: one, a concern with the curing-stripping stages, so that the sequence is worked back from there; or, two, that time is of the essence, so that sequence and processes are organised to minimise time taken. In most cases a combination of both will determine the outcome. In the first the sequence may be based on the technological demands of the building structure. For example, some sections of the concrete components can be stripped early, while on others the formwork has to remain in place for a long time. The dismantling sequence has to be carefully planned to ensure the release of the relevant falsework and the retention of that which remains.

When time is the priority then the system must be organised for fluidity in erection and dismantling. The sequences and processes are evaluated in terms of shortest possible times. With this approach it must be ensured that short cuts are not taken which jeopardise safety: the prime objective must be to minimise all risks to people. Because of the need to achieve fast times for the whole process it may be necessary to manufacture more sets of falsework than would be required if the pace of working were reduced.

In planning, the delivery of the falsework items should be incorporated in the sequence. Late delivery could result in inadequate materials being used, or the design being altered to suit those materials already on site. Either eventuality could lead to failure.

The Bragg report's list of duties for the Temporary Falsework Coordinator summarises this section. Is the design brief adequate and does it accord with actual site conditions? Has each element of the design been checked by a competent person, and has the falsework been considered as an integrated whole and approved by one person? Has the design been passed to the Engineer and his comments acted upon? Are the actual loads encountered on site, particularly live loads, no greater than those assumed by the designers? Is there a realistic programme for the delivery of materials to site? Have there been any changes in materials or construction? Are these significant? If so have they been referred to the designer and his approval obtained? Has each element of the falsework and the whole assembly been inspected, and faults rectified or alterations to design approved? Does the loading programme agreed on site accord with the designer's assumptions and intentions?

SUMMARY

The technological form of the building will determine the type and range of aids required for its production. The available aids for production will influence the technology of the building.

Eight principles for production were described; in considering any production process the answers to the questions posed by each principle will need to be formulated and integrated to form a coherent programme.

The changing nature of material (from raw to finished components) has implications for the aids used on site.

There is a continuing trend towards mechanisation of site processes, and the use of plant and equipment requires special attention. The characteristics of each item need to be ascertained so that its potential can be safely realised.

The construction of temporary works can become a production process in its own right, with programmes prepared, method statements issued and plant and equipment employed in their erection. There is a necessary emphasis on safety factors, as falsework can be potentially lethal. A process for ensuring adequate design and construction was outlined.

QUESTIONS

1. Show how a space-frame roof structure will differ from an in situ concrete roof in the aids for production it requires.
2. Select a piece of plant or machinery and evaluate its effect on a particular construction process.
3. Relate a building project within your own experience to the eight principles described in the text. Illustrate where the

project processes did not meet these criteria, or where the criteria were inadequate to meet the needs of the project's technology.

4. Describe how unprepared softwood is handled differently on site from a finished softwood component.
5. Analyse the erection of an in situ concrete floor on a multi-storey framed building and produce a safety system for the process.
6. Discuss the role, and relationship to personnel on and off site, of a Temporary Falsework Coordinator.

Part Three
THE ROLE OF BUILDING SERVICES

7. Introduction

The role of building services is closely related to that of building technology. Part Two outlined the major developments in building technology, a significant part of which were attributable to building services. In Roman times water was supplied to towns by aqueducts and underground pipes. This allowed large numbers of people to live together and form stable communities. By conducting water in conduits a degree of control over its cleanliness could be achieved, thereby ensuring its relative purity. The Romans exhibited a commonsense approach to the need for clean water, even though their scientific knowledge of how disease can be carried in water was virtually non-existent. It was also the Romans who developed the first recognisable form of central heating. Although confined to a small number of rich citizens this marked a significant development in the standard of services. After the collapse of the Roman Empire in Europe the ensuing cultures did not utilise the technique, let alone develop it further, until many hundreds of years later. So why did the Romans produce this system? It is impossible now to be sure, but a number of contributory factors can suggest a possible answer. The Romans themselves (that is, those native to Rome and its environs) considered that they were at the highest peak of civilisation and culture. They were aware of the Mediterranean civilisations that had preceded them, such as the Egyptians and Ancient Greeks, and knew about their advances in art, architecture and science. There was a strong motivation to do better than the preceding cultures, especially in art and application of science. The power of the state was reflected in its most visible asset – its buildings. The more the state was revered and grew in power the more the quality of building increased. The people in power demanded well-serviced, comfortable buildings. The citizens of Rome looked upon their leisure and entertainment pursuits as a high priority; to eat and enjoy long baths was a major part of life. In Rome such activities could be carried on throughout the year as the climate was not harsh. Internal building environments could easily be raised to the required temperature. Hot baths were valued for bodily cleansing and as a communal activity, perhaps this latter aspect being the most important; thus, efficient methods were devised to heat large volumes of water.

As the Empire spread northwards into harsher climes the Romans literally 'felt the cold'. There was a need for them to have adequate heating in their buildings. A problem not fully overcome was that the structural style was based on their native Mediterranean architecture. Doors were wide, windows numerous and open internal courtyards common. Unfortunately this building form was difficult to heat. As a partial answer to this the hypocaust system was evolved and adapted to the demands of the outlying settlements. The need for a form of central heating system arose from the sensibilities of the Romans and their desire to produce buildings which demonstrated a high level of civilisation and provided an environment within which they could enjoy their pleasures. The role of building services was clearcut: to uphold the status and comfort of the Roman citizen, whether official, soldier, merchant, farmer or commoner.

At present there is much evidence that services in buildings are acquiring greater importance than hitherto. One measure of this is the analysis of construction costs. Buildings housing some form of work activity (such as offices and factories) have a services capital budget ranging from 30% to 60% of total costs. A large proportion of a building's running costs are a direct result of using services. The obvious costs are those for electricity, water and, where relevant, gas. It is likely that services equipment will require more frequent and major maintenance than a building's fabric. The fabric is intended to last the life of a structure with the minimum of maintenance, whereas many service units comprise moving or wearable parts which will inevitably need renewing. Many householders in the UK will need to carry out some maintenance or upgrading of the services from time to time, whether to undertake a rewire or upgrade the sanitary fittings. New forms of heating systems are being developed which are more efficient, and to reduce running costs it may be worthwhile to replace existing systems: in spite of the initial capital cost subsequent savings in fuel costs may well produce a pay back period of only a few years.

As building services grow in importance, so the services

engineer's status is enhanced. His/her professional role is promoted by the Chartered Institution of Building Services Engineers. Many engineers are employed in a commercial role, that is, they design, specify, procure and instal services. Consultants do operate on a fee basis as designers and supervisors of services work, but at present they form only a small proportion of the numbers of chartered engineers.

In this section the modern need for building services will be investigated. What is this need based on? Is it common throughout the world? Discussion will centre on how appropriate services are married to user requirements; factors to be considered in the selection of services will be considered. In this industrial age a wide range of approaches to design is adopted. How do these affect the choice of services? Finally, issues arising from system performance will be raised, which in turn lead back to the basic question of the appropriateness of the services provided.

8. The Need for Services

There is a physiological need for human comfort, whether a person is inside a building or in the open air. The open air is virtually impossible to control, but when buildings are constructed the internal environment can be controlled. In buildings people carry out many activities, each requiring a different environment. Modern society also has needs, over and above those of basic comfort, for facilities which ease and extend human activities. In this industrial (or post-industrial?) age buildings provide comfort, amenity and agency: *comfort* in the control of heat, light, sound and security (in a physiological sense); *amenity* in terms of utilities (electricity, lifts) to ease activities;*agency* in allowing an extension to human activities (communications, sterile environments). To satisfy these three levels appropriate services have to be provided.

COMFORT

Comfort is achieved by regulating heat, light, sound and aesthetics to create an environment within which the desired activity can be happily carried out. An understanding of human physiology and how this is affected by the internal physical environment is basic to adequate design. Thermal comfort criteria, based on two sets of variables, will be used to illustrate a discussion on the issues raised by the need for services. The variables are as follows:

Personal variables

1. activity
2. clothing
3. age
4. sex.

Physical variables

1. air temperature
2. surface temperature
3. air movement
4. humidity.

The physical variables can be produced, altered and controlled by services, bearing in mind the four personal variables. The most important personal variable is 'activity', which has the greatest influence on the physical variables. Three illustrations demonstrate this. First, a person sitting watching television uses little or no energy. The body is at rest and is only generating enough calories to sustain internal temperatures. If the air temperature is around 18 degrees Celsius then a satisfactory level of comfort will be experienced by most people. Of course, this level will differ from person to person depending on physiological state, clothing, age and sex. Assuming that the person is comfortable at this air temperature their comfort may yet be disrupted by one of the other physical variables. A too frequent change of air may create air currents which feel chilling. High humidity may prevent the body efficiently disposing of its water vapour, resulting in moisture remaining on the skin. When these two occur simultaneously a person will experience a high level of discomfort. In the long term this can lead to health problems. If condensation is allowed to form on the cold surfaces of the walls mould growth can occur, itself a health hazard. The humidity in the air needs to be controlled, especially when using fuels which give off water vapour as a waste product.

A person working in a factory at a machine may use some energy and thereby create excess heat. To cope with this bodily production of heat the air temperature needs to be lower than if the body is at rest. If the naturally occurring temperature is 22 degrees Celsius, discomfort could be felt. Natural ventilation may reduce the air temperature by enabling a large number of air changes per hour. But this may cause further discomfort from excessive draughts or by bringing in noxious fumes from other manufacturing processes. These could create a health hazard, or at the least make working conditions uncomfortable, the worker's resulting distress leading to a loss in productivity. Clearly, there is a need for the environment to be controlled.

The third illustration concerns a building used for eating – a restaurant. Two adjacent areas will require very different environments. The worker in the kitchen will probably want the conditions to be cool and clean, whilst the diner will wish for a heated room. The process of cooking the food will produce excess heat and water vapour; as the cooks are constantly moving around they too will produce excess heat. In the dining room the people may experience fluctuations in their bodily temperature, which is inclined to rise after a meal. Does the ambient temperature need to reflect the before and after eating states?

People now generally wear less layers of clothing than in previous generations. On the other hand, there is an increase in the number of thermal underwear garments sold, possibly because of the rise in average age of populations following the increase in the number of older people. Or could this be due to the inability of systems to produce the desired internal environments? It may be that the high cost of fuel for heating is making people reduce temperature levels.

Human expectations have risen with respect to the provision of internal warmth in buildings. In the UK house buyers demand adequate means of heating and the preferred system is some form of central heating. Systems installed during the seventies still relied on one centrally positioned thermostat, but now the demand is for individually controlled rooms or areas. This enables the temperature to be accurately controlled in that area or the heat to be turned off completely when not required. Different temperatures can be produced in different rooms to suit the activity taking place in each. Technology has developed the thermostat so that it can work from individual heat emitters and the public has created a big enough market to bring the price within the reach of all. In parallel with developments in service appliances is the uprating of thermal resistance in structures. This was prompted by the realisation in the early 1970s that energy was finite and expensive. Consequently, legislation was brought in to reduce the amount of energy lost due to poor insulation of walls and roofs. It is likely that the 'U' value in the UK will be further amended to reduce heat loss. Although the price of energy is high the levels of comfort expected also remain high. At present there is much concern about the levels of humidity in buildings, especially dwellings, and it is likely that the installation of dehumidifying appliances will become commonplace. Office buildings now have forms of air conditioning installed even though in the UK the climate does not warrant its use except on a few days each year. Air conditioning is primarily used to lower air temperatures and humidity, and if conditions are uncomfortable for only a few short periods in the year does this justify the expense of installation and running? What price human comfort? There is no answer to this as it depends on values, which are based on many factors and differ from one person to another. But it is increasingly evident that any office building in London, whether refurbished or new, cannot be satisfactorily let unless it has air conditioning. There is also a practical reason for its use – to produce a suitable environment for the sensitive computers and communications equipment now used. Computers produce some heat and can be affected by extreme temperature ranges and by dusty air. Machines require a 'comfortable' environment, as do humans. Prospective office tenants want an environment which will satisfy both people and machines and which can be accurately controlled.

In many manufacturing processes a high level of control is needed to produce the right climatic conditions. Excess heat from machinery has to be gathered and disposed of. In the past this heat was either allowed to dissipate into the internal air or was vented into the outside air. The former practice meant that during naturally occurring high air temperatures the internal environment could become intolerable, affecting people and machinery alike. If the latter solution was adopted the heat lost was a waste of energy.

A further consideration arising from the venting of heat into the atmosphere is the long-term effects this might have on the earth's climate. Is the air temperature rising? If it is, will the ice-bound regions melt and cause floods? Is this the concern of the building technologist? Indeed it is, because it might affect the types of buildings required and especially their services. There is also the moral aspect, and as responsible members in society the repercussions of technological decisions must be appreciated.

The foregoing paragraphs have raised issues regarding the thermal comfort of people (and machines) now current in industrial societies. The need to control the air has been demonstrated, whether to heat, cool or control its humidity or to cleanse it of pollutants. As well as services being used to provide the means to these ends the building fabric and structure also play a part. Developments in the provision of services have been paralleled by construction systems that allow for predictable internal environments. Doors, windows and other openings can be effectively sealed to prevent any unwanted air changes. Walls and floors can be constructed to prevent heat loss. Sound can be prevented from leaving or entering space.

There are two other senses which can be affected by the built environment, namely sight and hearing. Because the building envelope predominantly consists of solid walls and floors, natural light is difficult to project to the interior. Windows can provide enough light during daylight hours for most human activities, but many activities require additional light, especially if some distance from a window. In a domestic situation reading and sewing can demand an increase in normal light levels. Office work demands a constant and clear level of light. Precision work on machines needs strong light to give clear vision. During dark night hours human activity does not cease, as it tended to do in societies not able to create artificial light. Which came first? – the human demand to extend activities into the dark hours or artificial light enabling activity to be continuous? A chicken and egg question which is now academic – but what will happen in the future?

The need for light is obvious, but recent research has given it a deeper psychological value. Some forms of human depression are more acute in the winter when natural light is scarcer than in summer. Giving patients doses of intense light over a period of days has been found to alleviate their depression. Generally, human efficiency and activity appear to be retarded in winter, and this could be due to the reduced quantity and level of light. But too much light in quantity and level disrupts human physiological systems, so a balance is required which will suit the prevailing activity, ranging from no light when sleeping to strong light when carrying out precision work. This balance is now achieved by using different types of lamps. Floodlights illuminate a sports field;

softer local light is used for reading. Artificial light can also produce different hues, ranging through the colour spectrum, to suit a given activity.

The problems of noise are only now being recognised, especially its effects on people's health (other than the obvious one of impairing hearing). It has been recognised for years that continuous exposure to high noise levels can lead to deafness. Perhaps the first examples of this were the mill workers at the time of the Industrial Revolution in the UK. The clothmaking machines had many moving parts, whose materials were such as to create a very high level of noise when in contact with each other; the machines were unenclosed; there were many of them on large floor areas; working hours were long, six days per week. Because of the noise the human voice was inaudible, so a sign and lipreading system evolved: this served the workers well outside the factory, as in later life many became deaf.

In recent years this problem has afflicted pop musicians and the habitués of venues offering loud music. Their hearing has been damaged. Should ear muffs be worn when listening to loud music? An improbable solution maybe, but what about the effect of the noise on those not involved? The general public is concerned about levels of noise, whether from music or machinery or aircraft. It is seen not only as possibly harming the ear but also as a nuisance which reduces the comfort and enjoyment of life. Consequently, legislation has been enacted which controls the levels and periods allowable. This legislation is policed by local authority environmental health departments. Private cases can be brought to curtail noise or to claim damages owing to noise nuisance.

Some types of noise pose an additional hazard to health and building stability by producing vibrations. All noise is a vibration but at many levels this is not a problem. Noise generated at lower levels, such as vehicle engines coupled with heavy loads, produce unacceptable vibrations. Structural damage can be caused if a building is subjected to a constant barrage. The vibration also obviously causes discomfort to people and in extreme cases can lead to health problems.

Noise has always been an offshoot of work activities on construction sites, especially as they have become increasingly mechanised. Excavating machines, concrete mixers, vibratory pokers, cutting tools and pneumatic hammers all produce excessive noise. This can harm the machine's users directly and, indirectly, those people in its proximity. The operators are expected to wear ear muffs; the machines should be 'jacketed' to reduce their noise output. Present-day machinery design must take into account the need to reduce the levels of noise generated at source.

The Building Regulations 1985 make specific mention of noise control in the structure of new and refurbished buildings. Airborne sound is to be reduced between the walls and floors of dwellings and impact sound controlled between separate users' floors.

So far this chapter has discussed the problems arising from the basic human need for comfort, with respect to heat, light and sound. As human development has progressed this basic need has been extended into a need to create services which ease activity. This will now be discussed under the heading of amenity.

AMENITY

Amenities are facilities which make life more agreeable and easier. They go further than just providing the basics of human comfort. For example, a moving stair escalator in a shop: stairs are an adequate method of moving from one level to another, but escalators make this easier.

Prosaic examples of amenity abound, some solely for the better control of services and some to prevent health problems. An example of the former is a timer control switch for a central heating system and an example of the latter is a bidet. Neither can be said to be absolutely necessary for life's functions but both contribute to making life easier and healthier.

Amenity provision is a selling factor in modern living. In most societies there is a belief that life can be improved by the acquisition and use of objects and processes. Whether or not this does lead to greater personal happiness is debatable. Some would say that to surround oneself with gadgets is pure self-indulgence and a wasteful luxury. Some services do contribute to making life easier but might be harmful if carried to extremes. For example, should people climb from floor to floor in shops, as the exercise will certainly do them no harm (whereas the lack of exercise certainly does) or should they use escalators? The escalator is certainly quicker and easier, especially when carrying shopping bags, and from the shop owner's point of view customers should be given the means to easy access to all floors so that they will not be deterred from buying the goods. In commercial terms the provision of escalators is highly desirable, enabling more people to get into and around the shop at a faster rate, thereby increasing the possibilities of sales and a subsequent increase in revenue.

There is a basic need for sound human health and this can be achieved by simple hygienic provisions, but modern societies demand a higher standard of cleanliness. It was for health reasons that building services were first initiated and developed. Water was drawn from streams, rivers, ponds and wells but as town populations grew these local sources could not provide enough at the right quality and quantity. The disposal of human effluent became a major problem in urban communities. Open drains in the street caused epidemics and at the best of times the smell must have been nauseating. There was a great public need for the provision of clean water and sanitary waste disposal, especially when disease was found to be related to the levels of waste hygiene and the purity of water. A point worth noting is that improvements in these services did not come about as a result of public pressure or commercial interest. Legislation from government had to be introduced to compel those people producing the built environment to provide drinking water and satisfactory means for waste disposal.

In a climate such as the UK's air conditioning might be said to be an amenity, but it is likely that its use will in future be seen as a necessity and become widespread.

AGENCY

Much of industrial human activity cannot be carried on without the aid of supporting facilities. These can make the activity comfortable; enable the range of activities to be extended; replace human endeavour entirely. The following illustrate each of these three aspects: a kitchen sink waste disposal system makes food disposal easier and perhaps more hygienic; mechanical ventilation allows work to take place in hot internal environments; automatic control systems for central heating switch on when set to required times.

The creation of sterile environments makes sensitive and complicated processes possible, for example chemical research or medical experiments. Indoor sports arenas allow activities such as badminton to be played in draught-free conditions and athletic events to be put on throughout the year. Computer rooms provide controlled atmospheres for the machines, together with their ancillary network of distribution services.

The third aspect is illustrated by the complete automation of factory processes. Machines, robots, computers, environment, building fabric, are totally integrated to create a system for the process. Warehouses are fully automated and controlled from a computer. The warehouseman requests the goods via a computer, which directs a retrieval machine to find the goods and take them to the loading area for onward transit.

As technology develops in manufacturing, communication and leisure industries the need for services to create the environment and enable the activity to proceed will increase. As wilder areas of the world are populated, albeit mainly in the pursuit of scientific investigation or for the gathering of fuels and minerals, buildings will have to provide protection and serve each activity. Oil and gas drilling and recovery platforms in the North Sea are there to do a job, but they must also provide acceptable living conditions for the operatives. The structure's prime task is to extract the fuel, but of course the other two aspects must be considered so that this can be done efficiently. Living and relaxation facilities are built to a very high standard and are geared to absorb the stress of living on a man-made island in a hostile environment.

SUMMARY

The need for services is widespread and increasing. The range and complexity of human activities demands environments to fulfil the expectations generated by them. Whether in resting, playing or working these expectations are for comfort, ease and agency. The boundaries between these three levels are blurred and what were once considered luxury facilities are now seen as basic comforts. Some of the more exotic provisions might be seen as frivolous, a waste of energy or completely unnecessary, but if the demand is there should the technologist satisfy it? Is it absolutely necessary to ensure that the roof of a concert hall is strong enough, and equipped to lift and hold, 30 tonnes of special effects apparatus for rock bands? In parts of India villagers are suffering because their local source of wood fuel has been virtually exhausted. A system for collecting methane gas from centrally deposited animal waste can be utilised, but it costs money which is not available. The technology is available, so why is it not put into widespread use? Should all people be enabled to enjoy the basic amenities before others extend theirs? Progress cannot be denied but the costs in resources must be considered. The ends must be appropriate to the needs and weighed against desires and expectations.

The quest for comfort, amenity and agency creates a demand for services and this need spreads through all levels and types of human activity. The factors which govern and constrain this need are presented in the next chapter.

QUESTIONS

1. Discuss the desire of society to extend the concept of comfort through to agency.
2. If buildings are becoming increasingly sophisticated in their provision of amenities, how will this affect the educational needs of builders?
3. Describe situations where the agency concept of the need for services dictates the built form.

9. Selection of Systems

Having identified a need for a service system the question arises of the means required to achieve the desired end. There are many factors affecting the choice of an optimum system, and these are listed below. No attempt has been made to put them in any order of importance, as what is crucial in one set of circumstances will have no importance in another. In any process of selection it is important to consider all the factors, even though some may be dismissed immediately as irrelevant.

Fuel/energy source

What form of energy does the service system need? Is it available? Does it need to be generated? Can it supply the necessary levels of power? Is there a need for a backup supply? Can a substitute fuel be used? What are the comparative costs of the energy? If waste products are formed how can these be controlled? Does the system minimise the use of energy? What is the community's overall policy regarding energy?

Building structure

The shape, size, height and layout of the building will affect the service system. Vertical and horizontal distribution have their particular problems, depending on the service. Many buildings are designed around the service provision, in other words the design of the services influences the design of the structure.

The types and materials of the fabric's construction will influence the choice of system. If a system is to provide thermal comfort the thermal conductivity of the external elements will influence the amount of heat to be generated and its mode of distribution. It might be better to enhance the standards of the elements (of course at a relatively higher cost) than to install a complete heating system.

Climate

The geographical position of the building will set the ambient climatic conditions. Service installations will need to meet those conditions, which can vary within relatively short distances. Building as a climatic barrier is discussed extensively in *Building Technology 2*.

Local, total or district

Are the service provisions to a part of a building (local); to the whole of a building (total); or to a number of buildings (district)? What are the economic factors relating to these three systems?

Component materials and methods of distribution

Are the service component materials appropriate, compatible and of similar durability to the fabric and finishes? Do the distribution materials harmonise with the aesthetics of the building? Are the distribution facilities hidden, disguised or made a feature of?

Position in building

Does the service system have to traverse between the substructure and the superstructure? Is it confined to one or the other? What are the relationship and connections with other services? How does it tie in with the major utilities provided by external authorities, such as the water board?

Loading/usage

Is the service to be utilised continuously, intermittently or in emergency only? Will there be peak periods of usage (for example, as with lifts to offices at the start of a working day)? Does the system require an overload bypass or trip mechanism?

An example of an emergency service is a sprinkler layout in case of fire. The loading will be at a maximum when called into use but hopefully will not be needed.

Even simple service systems are required to cope with an overload situation, for example the overflow pipe to a water storage tank. In complex sophisticated systems a series of bypasses may need to be installed. Another remedy to abusive overloading is to cut off the service completely. The choice of control is interlinked with the next factor, user participation.

User participation and control

To what extent do the users of the building require control over the services? Can they easily control them? Will individual control disrupt the system's efficiency? What is the degree of effort needed to effect control, by say, simple switch or adjustment to a number of items? Is it necessary for the users to understand the system? Do the users need to regulate the loading capacity? Can those parts of the system not required be easily isolated?

Standards

The question of standards relates to the value placed on the service by the users. But the users may well not be the specifiers and installers who initiate the standard. The installer's value criteria may be centred on initial capital cost considerations, or priorities in other parts of the building's elements. Therefore a basic service is deemed to be sufficient to meet the needs of the users, although the users would appreciate a higher standard. Which standard should prevail?

Safety

The safety standards of the services must be given a high priority. Failure should not cause injury to people or damage to the structure. In some cases the services are themselves safety measures, as in the case of sprinkler systems.

Installation criteria

Is the service to be installed in a new or existing building? Are there many individual systems, which require integration in design and installation? Can the system be easily installed, i.e. has it a high degree of buildability? The whole question of services installation is dealt with in Part Four.

Location of large plant

Many service facilities require items of large and/or heavy plant, for example, boilers as a heat source, or a substation for electricity. In large buildings, whether tall or expansive, plant may be needed in a number of locations to provide the desired capacity. The building structure will need to be modified to accommodate the heavier loads and to provide a suitable be positioned at optimum distances for distribution and accessibility.

Effects of services

The services themselves may generate unacceptable conditions which need controlling, such as noise and vibration. Plant rooms may have to be isolated from the main structure to minimise noise and vibration. Different types of services, such as water and electricity, will need to be separated from each other. The choice of service can be limited by its effect on the building, its occupants or the surrounding area.

Maintenance

Any service system will require some form of maintenance and the following questions will need to be answered. How often will maintenance need to be undertaken to the various parts? What degree of access is necessary? Who will carry out the maintenance? Will maintenance be based on preventive programmes or carried out on an emergency basis? Will the system need to be shut down, and if so what is the backup?

Obsolescence

Everything gets old and in the selection of service systems this should be considered carefully. What is the expected life of the system? Can it be easily updated when necessary? What provisions are made for its eventual total replacement? Is obsolescence 'designed in' so that as technology develops more suitable and efficient systems can be installed? Bearing in mind the finite nature of some energy sources and their variations in supply, to what extent is this relevant to new installations?

Energy costs of the building's materials

Consideration must be given to the overall expenditure of energy arising from the total input of materials and the subsequent running of the systems. The manufacture, fabrication and transportation of materials to the site incurs costs in energy: can some of this energy be saved by using alternative local materials/components? What amount of energy will be expended in the care and maintenance of the building and its services when components have to be replaced?

Legislation

The prevailing legislation, whether local or national, will have to be understood and applied to the design and installation of the services.

Life cycle costing/overall costs

A balance needs to be struck between the initial costs of the building, the installation of its services, and subsequent running costs. It might be appropriate to have high standards of fabric construction to reduce the need for services such as heating. Although initial costs will be higher than average, in the long term costs may well be reduced to below average. Techniques such as life cycle costing can indicate the comparative costs of alternative systems over their expected lives. Predictions regarding the frequency of maintenance and the costs of fuel etc. need to be made, as well as calculations about the present and future cost of money.

The cost of a building throughout its life stems from the technology used in its initial construction, so decisions made at the design stage will influence its overall cost and performance. Service installation is an integral part of the building both during and after construction and all the factors listed above will need to be considered for each building. No one system should be seen as a solution to a wide variety of service demands.

Summary

This chapter has highlighted many of the factors that need to be considered in the selection of a service system. They range from the technological constraints of the structure, through problems in assessing the optimum combination of service units, to the suitability of the system for the building's users, and costs in energy and money. A matrix of these factors, centred on the building, needs to be drawn up. In the first instance the building type will dictate the kinds of questions to be asked concerning the selection of services. After the initial priorities have been set, any of the other factors can influence the final choice. The selection process needs to take place against the backcloth of the contextual framework – specifically the factors of safety, economics, legislation, user values and function.

10. Design

Three alternative rationales for the design of the service elements of a building will demonstrate the range of approaches which can be adopted during the design stage. They depend on the perspective that the architect, or main building designer, adopts towards the overall look and function of the structure, fabric and finishes. Within that perspective the service systems must form an essential ingredient. In other words, the service design must be in accord with the prime design concept. Low impact technology relies on the building rather than the installations to meet the demands of comfort and amenity. The functional approach marries the 'state of the art' development of services provision to traditional (but modern) construction techniques. High technology sees the services as the basis of the building, and the shell and its structure primarily as vehicles for their efficient performance. These three approaches are now discussed in greater detail.

LOW IMPACT TECHNOLOGY

This rationale is sometimes referred to as 'soft' technology. A minimal use is made of appliances. In this context, 'appliances' are all those service units and installations which are in effect added to the building. To make and fit them involves work, money and energy, and to keep them running entails more of the same. This approach sees the solution to the basic comfort and amenity criteria in the building itself. The emphasis is on creating an internal environment which requires minimal use of services. For example, to control the amount of thermal gain from windows an air conditioning system can be used. The 'soft' option is to fit shading devices to the outsides of the windows. These can work automatically according to the intensity and position of the sun. Running costs can be low if operated by solar energy cells. The initial capital costs of the two options are similar, but the running and maintenance costs of the air conditioning will far exceed those of the shading system. A drawback to the shading device solution is that the aesthetics may not be acceptable. The building's appearance will be dominated by the shades, and internal shading devices may not be acceptable as this may reduce the amount of natural lighting. Artificial lighting might have to be switched on, creating extra costs.

The low technology solution is an extension to the building. It is part of its structure and fabric and, therefore, should in theory be considered alongside the design, build and use criteria. It thus falls within the province of the architect at the early stages of the design. While it is important for service design to be integrated at an early stage, whatever approach is taken, with this rationale the architect can exert greater control over the construction detail and finishes as these are part of the building's fabric. If an appliance is used, on the other hand, the architect has little control over its design. Many are mass-produced to achieve economic viability and can only be of a uniform appearance. The architect has to make a choice from a possibly limited number, none of which might fit ideally into the design. Another problem with appliances is that they are made to suit average or most common functional requirements. As most buildings are unique the services should, however, be related to particular sets of circumstances and performance requirements. The low technology approach seeks to limit the amount of energy (money) used in installing and running services, but still to maintain the highest levels of internal comfort and amenity possible with present technology solutions.

FUNCTIONAL TECHNOLOGY

This term embraces the majority of service designs and installations. Modern construction design is integrated with the developments in service components and each is made complementary to the other. A typical illustration of this approach is in new house building. Improvements and efficiencies continue in constructing the fabric of the building, notwithstanding the upgrading of building legislation regarding thermal, sound and fire insulation requirements. The fittings installed are also more efficient in their usage. Water boilers have been redesigned to improve fuel economy and provide water instantaneously. There is a far greater range of fittings available to be used in the house: from bidets, to showers, to waste disposal units in kitchen sinks. Different types of heat-producing appliances are becoming commonly available, such as heat exchangers. The fabric, structure and layout of the house will need to accommodate these new developments. An example of a situation where the structure of the house needs to meet different demands is in the need for a chimney. At one time all houses had a chimney flue designed to cope with coal fires. A change to smokeless fuel caused problems, especially in older chimneys, as the waste gases rose more slowly and deposited their chemicals on the side of the flue. These caused excess degradation of the bricks and mortar. With a trend towards gas-fired boilers and room heaters a totally new flue type was required. Precast concrete units were built into the wall. These were not structurally strong and when recessed into the wall did not add to its strength. The coal fire chimney was constructed and bonded into the wall, and the chimney breast with its supporting piers actually strengthened it. The wall with an older type of chimney is far stronger than the wall having a modern type of flue. It remains to be seen whether this is a significant weakness. In the case of a fire will all the walls collapse? The last wall to fall in older houses was the one with the chimney.

The process of this functional technology is dependent upon a mutual interchange of building design ideas, performance specifications and service units. The architect sets the scene and calls upon the services engineer (or unit supplier) to design (or supply) according to the design concept. Unless these two phases of the process are complementary the service installation can appear as an appendage to the internal aesthetics. Service runs may be obtrusive; may pass through structural members; have poor accessibility; or not be adequately protected. Performance criteria may not be met, resulting in inadequate service provision, leading to uncomfortable conditions (and possibly waste). Excessive maintenance could occur as the system is inappropriate for the purpose.

Most building owners and users are satisfied with the functional approach to meeting their service needs. It does produce buildings which reflect the current innovations in service units, and with care, knowledge and integration an efficient design can be created.

HIGH TECHNOLOGY

This approach to the problem of catering for the needs of building users views the service as an essential aspect of the design, around which the building is placed. The internal performance of the building is of paramount importance as it is likely to be housing complex, sensitive activities or to require large areas free from service intrusion. In the case of the Lloyds of London building both a sensitive communications system and large areas were required. To ensure that these large areas were adequately maintained at the correct air conditions a sophisticated service system was needed. The stairs, lifts, toilets and recirculation of air were placed on the outside of the building. Viewed from outside all these services, with their interconnecting pipes and trunking, can be seen. Pipes and ducts run horizontally and vertically; they are part of the architectural style of the building. This concern to demonstrate vividly the service elements extends to the roof, where brightly painted cranes are mounted. These support and control the cradles used for the maintenance of the façade.

As the service element becomes increasingly necessary to satisfy the needs of comfort, amenity and agency there is a corresponding increase in the physical presence of the installations. The high technology solution accepts the services as being an essential element of the whole building and they are featured in their own right. Bright colours highlight their position; bold metals give unequivocal shape and dimension; control units stand in commanding places. The designer of the services must not only satisfy performance criteria but must also blend the apparatus into the architectural style and detail. The services engineer must demonstrate an architectural awareness and be able to contribute to the overall building design. For example, when the lifts are placed on the outside of a building or internal wall, and encased in glass, the machinery and cage need to be incorporated in the architectural design. In more traditional buildings the architect only needs to approve the appearance of the lift doors and surround which are encased in a concrete shaft.

As most service installations are designed to high engineering standards and tolerances it follows that the elements into which they fit must also be to the same standards. Most service units are manufactured off site and assembled in place, therefore the 'place' must be to the correct dimensions. Holes for pipework or ducting must be correctly dimensioned and positioned and this information needs to be ascertained at design stage and entered on the working drawings. All the different designs need to be overlaid to ensure they will all fit into their designated areas and ducts. The high technology approach will place greater demands on the relationship between the separate services. Each, as they are left exposed, must satisfy the architectural style and also complement each other with regard to colour, shape and position. Does the larger duct go in front of the smaller? What brackets are to be used to reflect the architectural style? When the services were hidden these questions were not being asked, let alone answered. Adequate protection will be required, especially as they are not placed behind some other material. They will need to satisfy the following criteria: meeting the standards for safety; capability of removal for maintenance etc.; ability to cope with any possible impact; possession of surfaces which require the minimum of cleaning and blend in with each other and the decor.

Three approaches to service design have been described which exemplify the range of rationales. The distinctions between each are by their nature blurred and although some buildings can be clearly identified with one of the approaches, many encompass facets of two or all. Whatever approach is taken the design is still constrained by the physical aspects of the buildings. These are only briefly described here as they are considered in some depth in *Building Technology 2*.

PHYSICAL CONSTRAINTS ON SERVICE DESIGN

When a new structure is designed the services have to be fitted in at the drawing board stage. Some readjustments may need to take place to accommodate them, but for new work the task is made easier as the fabric design can be changed to suit the demands. When designing new services for existing buildings it may be impossible to alter the fabric, and then the services have to be made to fit. Unfortunately, this may result in a less than optimum solution being installed. The building form may govern the service provision.

The actual location of the building will also influence the service provision, and this in turn dictates the approach to design. If there is no suitable energy source available a low technology approach might be appropriate. The building's relationship with others and its effect on them as regards the service aspect (problems of noise and waste) will need to be taken into account. Fire precautions depend on the capacity to fight fires and provide adequate means of escape, and the building's location will influence this aspect.

In other words, the selection factors that were present in Chapter Three will all have to be considered and related to the building's form and location.

11. System Performance

The design process is undertaken to ascertain and forecast the appearance, fitting and performance of the building elements and components. In its purest sense service design is centred on the performance aspects. The installation is required to perform a function – to heat space; dispose of waste; move people vertically. In the final analysis the system will be judged on its ability to carry out its allotted task. One of the difficulties confronted by designers is to predict accurately how the system will perform, and the reader may well be able to recall instances of some form of system failure. Of course, many failures are due not to poor or inadequate design but to abuse by people or accident. Nevertheless, a little more attention to detail and a greater knowledge of use patterns might well have mitigated the failure.

A system can only perform to its initial design specification. For example, a lift installation to a college of higher education was designed for a level of usage based on the number of students in the building. Numbers are now in excess of that original level and as a consequence the lifts are overworked. This brings in its wake frequent breakdowns and the necessity to reduce the planned periods between maintenance visits, which in turn results in increased costs. The design brief was adequate initially but circumstances now render it inappropriate. Is the designer at fault? Could the increase in numbers have been foreseen? Would it have been economically justifiable to uprate the original lift provision to cater for a possible increase in demand? In any process of system evaluation like standards should be used. In other words, to be assessed relatively the performance of the lift in the foregoing example must be set against the initial criteria. In practice this is not possible because the extra usage cannot be curtailed: it is thus not possible to judge the performance accurately.

The manner in which a system performs is the outcome of its design. The selection criteria set out in Chapter Three should be fully considered in each case and appropriate solutions sought. If the optimum solution has not been found at the start then subsequent performance is bound not to meet expectations. A vigorous questioning process should be implemented and the answers tested and correlated against each other to ensure that the system is capable of meeting the performance requirements. If it is not, then the reasons should be stated and appropriate allowances incorporated in the design. Information about its limitations should be brought to the attention of the system's users.

THE EFFECT OF USERS ON PERFORMANCE

Most buildings and their installations go through the three stages of design, construction and use. The last stage can be quite remote in time and thought from the first. The user may have no indication of the designer's original ideas and little understanding of how the system works. To many people systems are there to be switched on, used and forgotten until they fail to perform. This lack of understanding cannot be wholly blamed on the user. Without being given the basic information no one can be expected to use any mechanical apparatus effectively. Even a simple operation such as flicking a switch may require prior knowledge, for example about the appropriate time to do this.

Operating a system at the wrong time may prove disastrous. The increase in degree of sophistication in service controls brings in its wake some interesting dilemmas. Take the case of a central heating system being able to switch itself on and off automatically according to the environmental conditions. Sensors can assess the prevailing air temperature; if this is above or below the desired set temperature they will send the relevant message to a switch which in turn will control the production of heat. The user has no need for knowledge of the process, the only task being to set the required temperature. Should the user have such knowledge? In an unsophisticated system such as an open fire the user usually had a full understanding of its operation and it could be visually overseen. Of course there are many drawbacks to this form of space heating and a return to its widespread use is not being advocated: the point of the comparison is that while the user may have full control when the central heating is working, control is lost when it malfunctions.

A further problem is that many types of systems tend to lose their efficiency over a period of time. Does the average person know when this loss in efficiency becomes costly? If the designers and installers do provide a detailed set of instructions and a fault diagnosis chart will that not create further problems, in ensuring safety, for instance? Is the average person capable of undertaking service checks or remedying simple faults? In some ways the user has gained greater control over certain aspects of service provision, but in other areas he has lost control. In large complex buildings there have to be permanent staff to ensure the efficient day-to-day running of service equipment. Many systems are automatic but still require some oversight, and staff must have some knowledge of their basic operation. They will constantly check the performance, but will call in experts when the problem is beyond their capabilities. The question for a building owner is: to what extent can these systems be overseen, bearing in mind the economic factors? What type of skills are required and at what point should additional skills be easily controlled and serviced by the building's user? Should the latter option be chosen instead of a theoretically efficient system which requires regular servicing and, if faulty, involves costly repairs?

Each building is unique and will therefore exhibit its own operational characteristics in the area of service facilities. The building's users will need to appreciate the nuances in operation and performance attributable to the singular nature of its particular environment. Buildings having similar, if not identical, service systems may require quite different operating procedures. Unfortunately these can only be ascertained in practice. This may produce some inefficient operations in the short term but if the mistakes are rectified and eliminated, and subsequent users are not allowed to repeat them, in the long term the services ought to become efficient.

It was mentioned earlier that there is usually a gap between designers and users in time, thought and communication. The initial designers may have little to do with the building after the defects liability period. Subsequent change of owners and/or users will render the gap virtually unbridgeable. So how does the design team know if the system created is working to the performance specification? In many instances the feedback will only come via major failures when the designer is called in to advise (or be sued for damages!) When small problems arise which are cured by the users the designer may not be involved at all, and therefore not be aware of them. This means that lessons cannot be learnt and incorporated into future designs. It could be that mistakes are repeated until such time as their accumulation is overwhelming and is eventually publicised. There are situations where direct feedback is possible but these are generally reliant on the client organisation's providing it. Such bodies as health authorities, commercial organisations with in-house designers and/or builders, and public housing authorities, ensure a constant feedback cycle. Initially the design may be worked up by the in-house technologists who will be party to the rationale and detail of the systems. These technologists will monitor the construction process and be on hand constantly whilst the system is operating. Any problems encountered by the users will be fed to the technologists who will in the first instance undertake remedial action and then note the problem. Subsequently any modification arising from the fault can be incorporated into the next system.

It is possible to give feedback to a wider audience by publishing papers, articles etc., but people are reluctant to do this because it means admitting to mistakes. Then again, will people read and take action arising from such publications?

Where the designer, supplier, installer and maintenance engineer are all within one organisation, direct feedback is possible. This occurs when the client has purchased a 'package'. The 'package' providers take full responsibility for all aspects of the service, so it is in their own best interest to eliminate problems which require unscheduled maintenance visits. The least number of visits under a service contract means maximum profit for the provider; alternatively, they can price their system at more competitive rates and therefore gain a larger volume of business.

System performance is determined by: initial design criteria; the specification which resulted from them; integration with the building structure; interrelationship with other systems; user characteristics; maintenance standards and frequency; and feedback to original designers. The combination of these factors will ultimately decide how the system will perform. For example, the system may be perfectly set up and working, but if maintenance levels are not sustained performance will deteriorate. It may be that the specification is inadequate and no amount of remedial work will improve performance. One system may be counterproductive to another, such as too many air changes demanding heat production beyond the capabilities of the prime heat source.

In the final analysis performance must satisfy the building's users and owners. The human activities need to be serviced, and if that is carried out effectively and economically then it can be said that the system is performing well.

12. Summary

This section, on the role of building services, has highlighted a number of issues relating to the need for services and their control and use. There is an undoubted need for services in buildings, whether the climate is warm and dry or otherwise. A basic comfort level needs to be achieved to sustain life, but the limit to which facilities producing these comforts, amenities and agencies can be developed has not yet been reached. Questions of appropriateness need to be addressed, such as: should domestic dwellings be fully air conditioned? What were considered as amenities a few years ago are now seen as basic comfort provision, for example central heating. As science and technology make advances service systems are required to enable and support these endeavours. Services become an agent in achieving further ends. The definition of comfort is widening to include aspects such as light and sound. Noise too can be a hazard to health and building structures. The services themselves can be a major source of noise, and suitable means must be devised to reduce this.

There are a number of factors which need to be considered when selecting service systems. They range from consideration of the contextual framework factors such as national energy policies and legislation, to the structure itself, the interrelationship between services, and finally, but not least, to how the systems' users will manage the total environment. Each building, whether new or existing, must be considered as a unique environment and all the selection factors will have to be addressed. This process does not mean that the wheel has to be reinvented every time a design is commenced, as much of the information is already available. It is quite possible, though, that feedback is not as effective as it could be, and so a consideration of all the factors is necessary to produce a worthwhile functional solution.

The manner in which a system is designed can affect its relationship to the rest of the building's elements and, therefore, influence the efficient functioning of the building. Three types of design rationale were described to give the range of approaches, these being low, functional, and high technology. Each approach makes an impression on the way in which service systems are perceived in relation to the building's function and its users' acknowledgement of their surroundings. The majority of buildings will fall within the rationale of functional technology, where the services are necessary to provide comfort, amenity and possibly agency.

Finally, the service system must be judged on its performance, and a cycle of design, installation, performance and feedback to design should be established. Performance itself is governed by many variables and can be affected if they are not in harmony. A poor initial design can never be made to perform correctly. Misuse by operators can limit the efficiency of a system, so should all be designed to eliminate human manipulation? Is this possible? There is a need to review service installations constantly, especially those that use energy, to ensure that they are meeting requirements and giving the best possible performance and value for money.

QUESTIONS

1. Discuss whether present day standards of service provision are adequate to meet the demands for comfort, amenity and agency.
2. Should service engineers be divorced from the commercial, profit-making aspects of design and installation?
3. Analyse a building and evaluate the need for and provision of its services. Can they be improved?

Part Four
BUILDING SERVICES INTEGRATION

13. Integration with design

In Chapter Ten a number of approaches to design were described. The classifications of low, functional and high technology delineated the range of perspectives that are adopted when considering the role that services play in the building product. Whatever the approach adopted the design will also be affected by the contractual and working relationships of the construction team. It is these two aspects – design perspective and contractual relationships – which are addressed here.

DESIGN APPROACH

Where the low technology concept is used as the basis for design, the building's services are seen as part of the structure and fabric, not as fixtures and fittings that are added later. The design is treated holistically: the building provides all the means for servicing the user's needs. The control of the design is with the architect, or the person commissioned to design the building. Thus all decisions regarding the means to create comfort and amenity stem from one source. In effect the integration of structure design and service design is total. The structure and fabric are designed to serve the majority of the needs of the users in terms of function and the optimum environment. Only where the elements cannot meet these requirements will the services be seen as appliances.

For instance, consider the design of the external walls. The architect will be concerned with: the appearance of the walls; their proportions; shape; texture; colour; the relationship of doors and windows. In addition the walls must satisfy the functional requirements of stability, strength, durability, thermal sound insulation, fire resistance and weather resistance. The low technology approach obliges the design to concentrate on satisfying functional requirements to the highest standards so as to minimise the need for appliances. Taking this to its logical conclusion means that a structure can be built which requires no additional sources of heating and environmental control. Experimental houses have been built which have walls (and the other elements) so constructed that they meet the functional requirements at the highest level. Thermal transmittance is virtually nil, the passage of sound is dramatically reduced, durability and weather resistance is substantial, and fire resistance is optimal. One such house, built in Sweden, has needed an internal heat source on only a few occasions during the coldest days of winter. The internal temperature is maintained by heat generated by the house occupants themselves. The degree of insulation to the walls allows no heat loss. A system of doors and lobbies prevents heat loss when opening doors to the external environment. Windows have triple glazing systems which again prevent heat loss but also reduce the passage of sound. Another example is the so-called autonomous house.

Such a house is designed to be as far as possible self-sufficient in energy requirements. The technology is concentrated on producing a structure which both prevents the possibility of energy waste and ensures that what energy is required is generated on site and stored for future use. The house is not isolated from the land it occupies. The ground below is exploited as an environment for storing energy by means of water tanks. (It can also be used as a heat source if heat exchangers are employed, although these are appliances.) The garden is used to produce basic foods. Human body waste is captured in a tank to enable the production of methane gas; this provides an energy fuel.

The low technology approach to building design is demonstrated exceedingly well at the Centre for Alternative Technology at Machynlleth, Wales. Here are examples of energy-producing devices which use wind and sun, together with organic methods of growing food and raising livestock. But more importantly, as far as building is concerned, there are examples of houses designed to reduce the direct service element. Some are experimental in that they are not in everyday use, but others are fully lived in and provide homes for families. Theory is put into practice and the outcome is carefully monitored to test the success, or otherwise, of the construction methods. The emphasis of design is on the construction of the elements, not on utilising service appliances. This is a further illustration of the design process being based on the structure, and there is little or no distinction between that and meeting the needs of the users. It cannot really be described as integrating building service design, since it is not perceived as being divorced from overall building design. This is not to say that knowledge and expertise derived from service design is not used. It is, but only as a base upon which different solutions can be founded. Much experience has been gained from designing and installing service appliances, especially with respect to levels of efficiency. This information is used to underpin low technology designs and to give a basis for comparison.

The examples of the Swedish house, the autonomous house and the houses at the Centre for Alternative Technology show service design not being separated from overall building design. The design details and process are within the control of the initial designer, although he will draw on others' experience and information. One important aspect is that any innovations in the means of improving and developing the comfort and amenity levels are solely in the hands of the building designer. Any additional installations are subservient to the initial design and must conform to the main objectives. This situation is not so for the functional technology approach.

Functional technological concepts rely on the manufacturers and suppliers of service units to provide the means of creating comfort, amenity and agency. The design process must 'fit-in' the available appliances against the basic design criteria. The appliances and construction of the elements must be matched to each other. The designer is constrained by the possibility that the appliances will not match exactly his performance criteria, and has to rely upon the manufacturers and designers of service equipment both to be up to date in meeting the demands of users, and to be honest in describing the capabilities of their installations. The designer has to take the manufacturer's word that a specified unit will meet the performance requirements.

In the process of matching the appliance to the building structure the designer must have to hand all its physical dimensions and installing procedures. This information needs to be incorporated into the overall design at an early stage in order that the structure can accommodate it. Unfortunately there are many instances where this has not been achieved. Reasons for this are: the environment specification was not thoroughly thought through and confirmed early in the design process; the choice of appliance was left to the latter stages of the pre-tender/estimate stage; the appliances were modified between the specification and installation stages; a cheaper appliance became available after the original choice, requiring changes in dimensions and fixing procedures; the client changed the environmental requirements. At least, with this design approach, if a change has to be made its repercussions on the structure and fabric of the building will be relatively small. This is because the basic elements were designed independently from the service appliances; they stand in their own right in meeting the functional performance requirements. Some minor design changes (sizes, layouts, etc.) may be required to accommodate the change but these should cause minimal alteration to the basic design.

The building can be designed as a shell into which any appropriate and available service installations can be fitted. For example, some factories are designed and built in advance of any known user. The basic services are supplied, but allowance is made for any additional services required to meet the particular demands of the users. This will also provide for updating of the services, as they can be fitted in and taken out quite easily if they are not an integral part of the structure.

The functional technological approach is the one most commonly used in designing buildings, and for the majority of cases is ideal. The expertise of service designers and manufacturers is employed to the full and any improvements can be quickly incorporated. The building designer need not have detailed knowledge of the workings of the service systems and can, hopefully, rely upon them meeting the specified requirements. The designers are in a partnership that seeks to meet the users' demands. The strengths of each area of expertise are drawn upon, and as long as both are aware of the constraints within which each has to work worthwhile solutions will result.

High technology demands a correspondingly extensive knowledge of service systems on the part of the building designer. It is similar to the low technology approach in that it gives aesthetic and dimensional control to the builder or the designer. The major difference is that the service appliances become a dominant aspect of the building's design. They are there to create comfort, amenity and agency and as the building's function is to provide these, so the services must be fully recognised and acknowledged. Therefore they are not disguised or hidden, but are given prominence, whether as pipes, cables, ducts or fittings.

The building shell complements the service requirements, and the design process commences from the determination of the environmental conditions, levels of amenity and scope of agency. Services are fully integrated into the structure at the start of the design activity; indeed it could be said that the building's structure is integrated into the service design.

As service units and distribution networks are designed on engineering principles the levels of accuracy and tolerances are to a higher standard than those normal in traditional building. It is possible to design services to minimal tolerances, as the materials used are capable of exact dimensions. The architect can utilise these levels of accuracy in the design of the structure and fabric, so ensuring a high level of structural integrity. To achieve this a knowledge of materials, especially metals, is required. Traditional materials such as concrete and brick are difficult to form to exact dimensions, so problems could arise when trying to integrate dissimilar tolerances. A further problem in using concrete and brick extensively with exposed ducts, pipes, facilities etc. is that of producing compatible joints. As the materials have different properties it is difficult to design joints able to cope with them. Precision engineering techniques are adopted in the design of high technology buildings and the architect has to understand the degree of accuracy which can be obtained. An added complication in the attainment of correct levels of accuracy in the building's elements occurs where the majority of units are manufactured off-site. Factory production enables very high levels of accuracy to be obtained, but since the units are prefabricated by, possibly, many disparate manufacturers the overall integration needs to be carefully planned. The design process must first produce the individual service components that meet the comfort, amenity and agency brief, secondly interrelate these into a composite system, and finally create the building to give the basic functional requirements. The need for assembly drawings is paramount, and the drawing presentation system described below can help the designer ensure a conceptual integration not only of the building's services but of all its elements and components.

Co-ordination of working drawings

Daltry and Crawshaw (1973) showed that many problems arose as a result of technical information being poorly presented on drawings. Their research was followed up by Crawshaw (1976) to investigate how different professionals, such as service engineers, ensured that their installations could be fitted into the building. His paper concluded:

> This study has shown how easily uncoordinated drawings can become the major reason for difficulties created by inadequate information. Isolated coordination faults can be individually expensive and disruptive; taken as a whole they can have serious effects on the organisation and management of a project. There is, therefore, a general need for recognition that the coordination of designers' information output is an important task requiring an organised strategy.

His recommendations are briefly summarised below.

1. Systematic approach to drawings

Each working drawing should have a clear and specific purpose, and this is best achieved by a systematic approach to the production and arrangement of working drawings. One way of organising drawings is given in the BRE Digest 172, *Working Drawings*. The relationship between the drawings is shown in Fig. 4.1. The set should have a systematic structure comprising separate groups of location, schedule, assembly and component drawings. Drawing sizes should conform to the international 'A' dimensions. Location drawings are generally best presented on A1 size, at scales of 1:50 or 1:100 to prevent fragmentation of overall plans and elevations. References should lead the user directly to individual sheets. To aid the search for drawings, each should also have a short title in addition to the reference. The use of grid or reference lines to fix the position of views or parts of the building is recommended. A brief guide explaining the drawing arrangement should accompany the set, especially for those working on site.

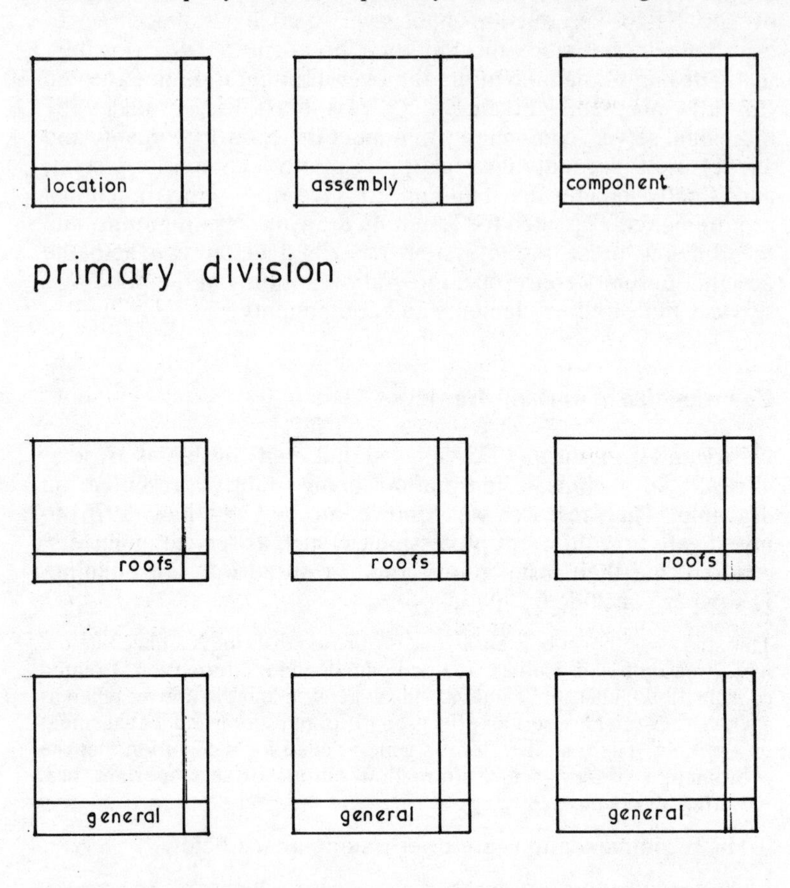

4.1 Structured sets of drawings

2. Combined service drawings
Service engineers and architects should develop a combined service drawings approach. At least one master drawing should show all the services, and their relationship to the building elements.

3. Design team meetings
It was found that design team meetings are often ineffective in dealing with coordination. An aid, such as a combined service drawing, will enable each designer to be aware of the consequences of their designs. Although the formal responsibility for

design is usually the architect's, all design team members must be prepared to accept some informal responsibility. With the use of aids and an acceptance of some measure of responsibility, design team meetings can be much more effective.

4. Unusual features
The purpose of the drawing should be clearly stated, especially when passed between designers. Any unusual features, such as an especially deep beam or unusual door swing, should be highlighted.

5. Information status
All data on a drawing appear to have the same status, whereas some may have a high status and others be there only as points of information. Any such 'soft' information should be identified as such, and the drawings themselves given a status identification.

6. Consultant involvement
Service consultants should be involved in the design of a building as early as possible. If consultants are presented with a completely designed building into which they are expected to 'fit' a service system, the opportunity to design out problems in coordination is lost.

7. Checking
Over the years more effective checking of drawings has been urged. Since this is an arduous task the adoption of a systematic approach to drawings might prove to be a time saver, as each drawing will have a clear purpose and checking will be easier.

8. Amendments
Unfortunately drawing amendments are often not controlled effectively. It is best to channel all amendments through one person and to keep a sub-group of drawings constantly up to date. It is likely that not all information on all drawings will be current, and the service designed must therefore always refer to the latest architect's drawing.

9. Copy-negatives
Copy-negatives have two main uses: passing information between designers and acting as base drawings. These should be produced as early as possible in the design process. Some information on the copy-negatives may be out of date so care must be taken.

10. Scales
Coordination is very difficult if each designer prepares drawings having different scales. A decision on a common scale for, say, the plans, must be agreed before production of drawings commences.

11. Site staff
Site staff must be well versed in the system of drawings used, so that the design information can be effectively translated into a three-dimensional building with all its facilities.

Many of the above recommendations can be met by using the

high technology approach. In essence this gives full recognition to the need to integrate and coordinate services – the main task of the architect. Whether or not systematic drawings are used is not so important here since they are only a mechanism through which coordination can be realised. If this realisation is the basic design rationale then, in theory, any form of drawing presentation will be adequate, but bearing in mind the conclusions of BRE CP 181/73 there might well occur major differences even in drawings which have been prepared with particular care. The lack of guidance generally about what is to be communicated – the information needed to build – is the underlying cause of many deficiencies in working drawings.

The use of a systematic approach to achieve coordination is essential when adopting the functional technology approach. As there is a distinct division between the designer of the building and the designer of the services, the use of master drawings assists in bringing the two designs together. It might well be necessary to produce separate assembly drawings for electrical wiring, waste pipe runs, water supply, heating distribution etc., each showing the pertinent fixings and relationship to the structure. Also necessary will be an assembly drawing which shows all of these at relevant places in the building, i.e. where they come together or intersect. It should be the responsibility of one person to produce this drawing and to pass it to all the service designers so that they are fully aware of all the other service positions.

The low technology approach to service design and installation will investigate the problems of coordination. As with high technology, the control of service design lies primarily with the architect. Since the use of appliances is considered as a last resort in providing the required levels of comfort and amenity, their integration is less of a problem. The building structure and fabric itself provide the means to create the internal environment and, consequently, the service need is naturally coordinated into the actual construction. Nevertheless, problems could still arise regarding the integration of services, especially if the structural solution is inadequate to meet performance, and the appliances (with their attendant ducts, cables etc.) have to be fitted into the design at a late stage. Again, the use of a systematic set of drawings based on location, schedule, assembly and component, will enable the design to be modified relatively easily. For example, separate drawings can be produced to give the component (service unit) details and show how it is fixed into the building (assembly drawing). These would be additional to the set. Alterations will need to be made to the location drawings, but these may only be additions. It will probably be necessary to produce master assembly drawings to show the relationship of the service installations to each other and the structure. If the traditional drawing format has been used, i.e. plans, elevations, sections and details, then major revisions or even redrawing may be necessary to show the new installations.

In summarising the discussion on design approach two points will be emphasised. Firstly, whether a low, functional or high technology approach is used, the designers must accept and apply the principle that integration and coordination are vitally important design aspects. Secondly, by using a systematic approach to the production and presentation of design drawings, successful coordination can be achieved.

CONTRACTUAL ARRANGEMENTS

Four main forms of contractual arrangement will now be briefly described. Each will be evaluated against the most common design approach, functional technology, and some conclusions will be drawn. Finally, a general discussion will look at some issues affecting the low and high technology approaches with respect to contractual arrangements.

The four main types of contractual arrangements that are described here are:

- traditional (design/tender/main contractor/sub contractor);
- design/build (majority of design and production from one organisation);
- management fee (conceptual design/tender/negotiated packages for specialist work);
- client own (in house design/estimate/build/sub-contractors).

Traditional

A client engages an architect to design, produce working drawings, select any nominated sub-contractors or suppliers, invite a number of main contractors to tender, accept a tender, supervise works and agree payments to contractors. The contract is made between main contractor and client. Another contract is made between client and architect. The architect will advise his client on the need for any specialist consultants or designers, such as quantity surveyors or service designers.

Design/build

The client enters into a contract with an organisation which will design, manufacture and build. Usually these organisations are commercial companies promoting a particular system of building.

This enables production of similar elements which can be put together to give a variety of overall shapes and sizes. The design is the responsibility of the organisation, although an architect may be engaged by the client to advise and oversee the work. All specialist consultants are normally under the control of the design/build organisation.

Management fee

With this arrangement it is not unusual for the client to engage a management fee contractor before an architect. The management fee contractor will then advise the client as to a suitable architect for the building type. The converse can also apply, where the architect is engaged initially and, owing to the nature of the project, advises the client that a management fee arrangement is most appropriate. Normally, the construction work is let as packages, based on particular trades or elements of the building. For example: the frame; plastering; lifts; plumbing; floors; communications systems; painting and decorating; general builder's work. Each enters into a separate contract with the client, and – here is where a departure from the norm produces a very different relationship – the 'package' contractor may well be responsible for the detailed design as well as supply and installation. The architect will have set out a conceptual design framework and delineated its parameters, or have given the performance criteria. Within these constraints the contractor will have to produce the detailed design (for approval) and then supply and instal. This entails a very high level of responsibility for the contractor which, as he is a specialist with a wealth of experience, he should be able to meet successfully. As it is likely there will be many disparate designs the task of integration and coordination will fall to the architect, working with the management fee contractor. If the initial ideas came from a management fee contractor then he will undertake the task of coordinating both design and installation.

Client own

Some public and private building providers (who build primarily for their own use) undertake the design and build process using their own employees, supplemented where required by specialists. No formal contract is made between designer and client, or builder and client, but formal contracts will be made between client and outside specialists. In this arrangement clear lines of responsibility should be drawn so that designers and builders know how far they extend. The client can exercise direct control over the design.

The above four contractual arrangements, commencing with the traditional, will now be considered in relation to the functional technology design approach.

1. *The traditional approach* places the responsibility for design, and decisions for the choice of service systems, upon the architect via working drawings, a specification and perhaps a bill of quantities: the major choices are made at precontract stage. The architect determines the types and quality of the services appliances, based on previous experience and/or consultations. If the building requires many and complex services the architect will delegate the design responsibility to a services engineer. This engineer may be an independent consultant and therefore not restricted in the choice of appliances, etc. Alternatively, the appointed engineers could be a commercial organisation which will design, procure and/or fabricate and instal the systems. This arrangement will restrict the range of options regarding the service appliances, as the organisation will tend to favour its established suppliers and manufacturers.

As the builder will not have been selected at the design stage, the integration of the service installations can only be projected by the architect and service engineer – only two-thirds of the whole team. The design will not fully consider the methods and sequence of construction that will be adopted by the builder to achieve efficiency, speed and buildability. Inevitably there will be some disharmony between the methods prescribed by the design team and by the builder. Though this will not necessarily be detrimental to the construction outcome, it could cause minor communication and physical integration problems. The builder is presented with firm choices and it is likely that some of the services may be nominated under the contract, e.g. the builder must work with the specified supplier/manufacturer/installer. The main contractor will have to coordinate the nominated contractors and ensure their effective integration into the construction sequence. During the tender preparation stage the builder has to assume that the nominated contractor can meet the programme requirements upon which the total estimate is based. At this stage it is unlikely that the builder will firmly ascertain such issues as:

- lead time from placing an order to work starting on site;
- length of time actually required on site by the nominated contractor to carry out the work (coupled with the desired sequence of installation);
- detailed requirements relating to attendance and general builders' work in preparation for the installation.

It is not until the main contractor has been awarded the contract that these issues will be fully investigated and resolved.

In traditional contractual arrangements the majority of design decisions relating to services (together with all the other building elements) are taken and firmly established before the builder is chosen. The architect, and the specialist consultants, specify the installations and may even nominate the particular supplier/installer. The service system may well be integrated into the design drawings, but not necessarily into the construction programme and method of work.

2. With a *design/build arrangement* the designer, specifier and builder are within one organisation. Commonly, this organisation will have developed a system of building based on particular elemental structural units, floors and claddings. Components and finishes may also be available from a select range, together with the basic service facilities. A client will buy the whole building and the design and construction will be included in one price. The client may require specific finishes, components or services, and the cost of these variations will be negotiated with the design/build organisation.

The design/build firm usually employs directly the designers and building management team. In many instances they are accommodated in the same office/building, with site management located on the site. Thus the designers can consult directly with the manager/technologists regarding methods and sequence of construction. Direct feedback from site is possible regarding the buildability of elements and components, so that improvements can be incorporated into future designs. In large organisations specialist designers such as service engineers are directly employed. Again, the advantage is that they will be able to communicate directly with the main building designer and ensure adequate integration. Wherever possible standard designs, materials and construction methods will be utilised, based on past experience and present innovations. There is a danger that only tried and tested installations will be used which will not reflect the advances in efficiency of some service units. But as the design/build firm will be competing in an active market (there are a number of such firms in the UK) this should ensure up-to-date installations.

The production of the design drawings can be to a common format, as they all emanate from a single source. This format will be familiar to the construction management team, so aiding information retrieval – and interpretation. Standard drawings may be carried forward from building to building, especially structural details, again by their familiarity helping the site construction team.

Within the firm, procedures can be established which will ensure that:

- integration of all the building's elements, components, services and finishes is achieved;
- close collaboration takes place between the design team and construction team;
- note is taken of service innovations and they are incorporated where appropriate;
- feedback from site and the building's users gets to the designers and is acted upon.

If these conditions are met the technology of service integration and installations is likely to benefit greatly. The design will be feasible, it will take into account the user's comments and recent innovations, and efficient construction work can take place on site. With design/build there is a ready-made organisation structure which can ensure effective integration.

3. Recently, *management fee contractual arrangements* have begun to be used extensively. The method is not suitable for all forms of building and its implementation is influenced by the building's structure, services etc. This arrangement, which has many variations – not to be developed here – generally places responsibility for detailed design, such as services installations, upon the contractors who quote initially and are awarded the contract after a normal selective tendering procedure. It is common for the building's elements, such as frame, floors, walls, lifts, sanitary ware and plumbing, etc. to be let as separate packages. The architect will have produced a contractual design for the whole of the building; some items may be fully specified, leaving the contractor little scope for detailed design, but others may be described in performance terms, leaving the contractor to design and specify in full. This design is then approved by the architect and the management fee company. On this basis two criteria can be met: firstly, that the detailed design fits into the conceptual design and the installation is integrated aesthetically and functionally; secondly, that the contractor's proposals fit in with the construction methods and sequence of work.

What are the advantages of this contractual arrangement?

1. Specialist contractors can fully employ their knowledge in providing suitable solutions in design and installation.
2. The building's management team can oversee and firmly control the contractual arrangements and clauses, and the pre-contract activities of the specialist contractors.
3. Using a competitive selective tendering system for the individual packages, value for money can be obtained by the client.
4. The architect is freed from the production of detailed working drawings so that he can concentrate on the overall design and coordination of all the elements.
5. It enables a 'fast-track' method of construction, where work on

site runs immediately behind detailed design. This means that the pre-contract phase is reduced, as a large proportion of the detailed design is carried out during the construction phase.
6. The management fee company can directly control the specialist contractors, thereby ensuring efficient on-site integration both in programming/sequencing and in the physical incorporation of the structure.
7. Responsibility for the design and installation of a particular building's elements, components, etc. (as identified by the work packages) can be clearly defined and is usually fully attributed to the specialist contractor.
8. It can lead to specialist contractors introducing innovations and new methods and improving upon initial design concepts.
9. The specialist contractors can state clearly their requirements regarding fixings, builder's work and any temporary works, as well as influencing the method and sequence of their installations.

There are some disadvantages to the management fee contractual arrangement. For example, not all specialist contractors are able to produce detailed designs; although they may have the knowledge they do not employ people with design skills. Again, during the tender stage each contractor needs to produce a design in order to quote for the work. If, say, five contractors are quoting then five designs are produced, compared to one when using the traditional contractual arrangement. The architect, too, loses both absolute control over the detailed design and his supervisory role over the project. No standard or common bills of quantity are produced, which can lead to wide variations in price, design and specification; this makes the task of comparing one contractor with another a little more difficult. Finally, the task of integrating and co-ordinating both design and construction demands management skills of the highest order.

Under management fee, technological advances are more likely to be introduced into construction sooner rather than later. Generally, such advances are produced by specialist contractors, and as they are encouraged to produce their own designs the innovations can be incorporated into the project; they do not have to be assessed and then sold to an architect who might in turn use it in a future design and a traditional specification. In other words, the process of introducing new developments is short-circuited by the management fee arrangement.

In summary, the management fee method can allow specialist services consultants or contractors to produce designs based on their probable extensive practice and experience, and also to introduce new developments and influence the installation processes. In the final analysis the client or building user should benefit from the full application of this specialist knowledge.

4. Using the *client own* approach, some large private commercial firms (and public authorities) have developed in-house design and supervisory teams. The team may extend into actual site management, but for medium- to large-scale projects this responsibility is taken by the contractor. The building's design is developed by the organisation's employees, with outside specialist consultants brought in where necessary. With this arrangement the client can have direct control over the design, both in its content and its process. It should ensure a complete match between users' requirements and the building's facilities. In most cases, the buildings are constructed for the client's own use, for example factories for large manufacturing companies, and hospitals (or parts thereof) for regional hospital boards.

As the design is fully in the control of the client it is possible to be exact about the requirements. Design decisions can be easily questioned by the client. Designers will have an intimate knowledge of the users' requirements as they are in close contact with the client's main activity, whether it is producing other goods or providing a service. With respect to service systems, it is expected that these can be identified and incorporated into the design at a very early stage. Indeed, it may be that the client's prime activity is a manufacturing process which depends totally on adequate service provision. Two factors emerge from this situation: one, that the service designers will be totally familiar with the manufacturing processes' demands, and two, that the building (and its services) will be constructed primarily to accommodate the client's prime business activity of product manufacture. In this contractual arrangement the client's knowledge of his needs should be clearcut and detail design and/or other requirements can be effectively communicated to specialist consultants and ultimately to service unit installers.

How do each of the four contractual arrangements described above compare with each other? In practice, each has its merits and demerits, and a definitive comparison is impossible to make as one will be chosen in each case to suit the circumstances prevailing at the design brief stage, such as: general economic factors, legislation, the client's wishes, the advice he receives from the built environment professionals, the situation with regard to availability of design and build skills, time available for design and construction, and economic factors. Despite the difficulty of precise comparison, however, a few comments can be made with respect to the integration of building services in design.

The traditional arrangement will tend to provide the least favourable framework for effective integration, the best probably being the client's own team, with design/build and management fee in between. The traditional approach relies upon the architect to coordinate all the specialist design contributions (a difficult task): the specialists work to a specification provided by the architect which may not match with their own particular skills and processes; innovations and advances may not be integrated effectively; barriers can arise between the various designers owing to lack of clear communications. The end result is that the technological solution, although competent, may not be the optimum. Where, on the other hand, the client is fully cognisant of the design, there is greater probability of achieving an optimum solution.

Where time is the crucial factor the management fee system can produce the most effective design/construction integration. The detailed design is left to the specialists, who carry it out both prior to and during the construction process. They will have the responsibility, in conjunction with the management contracting team to ensure optimum design and installation integration. This arrangement does mean that communication must be of the highest clarity, and that is the prime responsibility of the management team.

The main advantage of design/build is that the client will know what is being bought at the early stages of the building process. The designer/builder will be selling a system which can be clearly described in concept and detail, through drawings and specifications with, possibly, already built and functioning structures as examples. Components should be standard and available. The time between inception and handover should be relatively short when compared to traditional processes, as most of the design will already be complete, and the production of the structural elements and components can commence quickly. Services units and components should also be readily available. A disadvantage of design/build is the limited range of alternative construction details. Any variation from the system will involve redesign and alterations in specification, with probable knock-on effects for other elements or components. These may also incur extra costs. There is a danger of incompatibility if the client demands a service installation not designed with the particular system in mind. If specific service requirements are needed by the client one of the other contractual arrangements would be better suited. The services can then be designed in conjunction with the structure and fabric, and each adapted to reach the optimum.

The issue of responsibility for design (ultimately the performance of the product) has already been raised in the description of the management fee arrangement. This is now explored further.

The traditional role of the architect has been to determine and design the majority of the building's elements, structure and components. Consultants will have been employed to produce detailed designs, but overall responsibility resides with the architect. There is limited opportunity for alternative, and perhaps innovative, designs to be considered. In the management fee arrangement a number of specialists can be asked to tender for the work package, each providing their own design solution to meet the overall brief. The management team can select the one deemed to be the most appropriate to meet: the needs of the client; the building's structure; the integration of design with construction; the user's activities; and, of course, cost. The responsibility for the effective performance of the design lies with the specialist contractor. His expertise is utilised to the full, and is more likely to take account of advances in technology. A criticism of this approach is that it is difficult to appraise a number of different solutions, especially as they are likely to vary widely in technological solutions and cost. Like is not being compared to like. With traditional drawings, a specification and perhaps a bill of quantities, contractors are competing on the same basis. But the question of whether or not the basis is the right one cannot be put by specialists, who may be able to offer more suitable solutions than other specialists. This is especially so in the field of services engineering, where many companies offer a comprehensive design and install capability. If the specification ties them to a preconceived solution their contribution to design (based on their detailed experience) cannot be included: they will be expected to take responsibility for the design, but are denied the authority to produce what might be a better solution, to the client's benefit.

The contractual arrangements, and subsequent relationships between design (and build) professionals, can affect the technology used in the building. Each building should be regarded as unique and the most appropriate contractual arrangements used to satisfy the prevailing economic/social/political/legislative/user values, the function of the building, and the clients' wishes.

SUMMARY

The three approaches to design – low, functional and high – do have an influence on the integration of design. Low technology assesses the building as a whole and seeks to reduce the input of service appliances. Integration is not seen as a problem because there is little to integrate: the building is designed as a total concept. Functional technology also requires a high degree of integration as many specialists produce separate designs which need to be matched and integrated with each other.

High technology considers the building primarily from the point of view of its service provision, and the structure is, in essence, designed to accommodate the services. The structure is integrated into the function of the building, with service design taking the lead.

The four contractual arrangements described also influence the scope for design integration. Traditional arrangements pose the most problems in achieving integration and coordination, whilst the client's own solution offers the best possibility of meeting the desired level of integration of overall design with, ultimately, specified performance.

QUESTIONS

1. What inpact would a predominantly low technology design approach have on the heating and ventilating industry?
2. Describe how a high technology approach can meet user demands for sophisticated communications systems in buildings, by those working in the financial markets, for example.
3. In selecting an appropriate contractual arrangement a problem arises when the client is not aware of the alternatives. Discuss how the client can be best informed.
4. Integration of specialist designers is likely to become more complex as specialist fields increase. Discuss how this might influence the contractual arrangements.

14. Integration with Construction

INTRODUCTION

This chapter explores the issues involved in integrating services design and installation with the actual work of constructing the building. There are two aspects of this integration. One, that the sequencing of the service systems has to be coordinated with the overall construction programme; and two, that the service installation has to physically fit, and be compatible with, the structure, fabric and finishes. As the final part of the sequencing programme, the task of testing and commissioning is the summation of the construction process. Fig. 4.2 shows the interrelationship between design, installation and commissioning, with the final link of feedback. Impinging upon this process is the nature of the building – its structure, fabric and elements; its functional use; and, as seen in the previous chapter, the approaches to design and contractual arrangements employed.

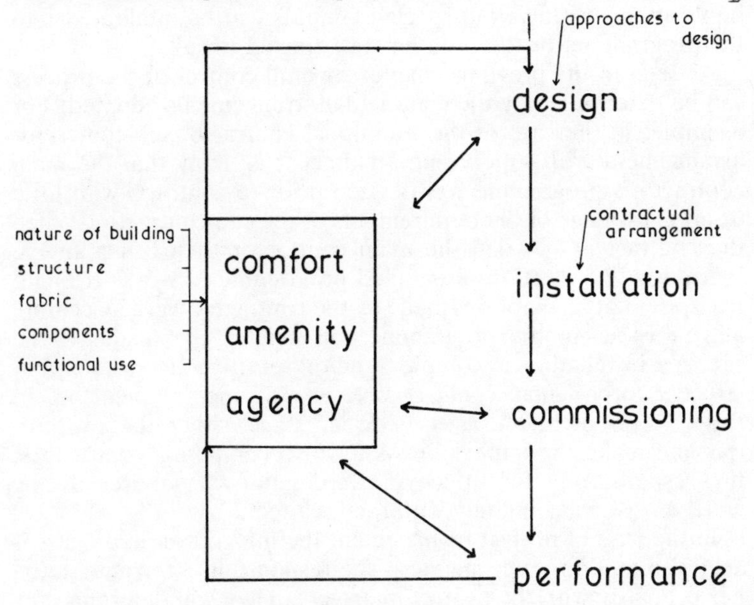

4.2 Integration with construction

INTEGRATION WITH PROCESS

The first aspect, sequencing of the services installation, is dependent upon: the contractual arrangements; the degree of specialist involvement; where the responsibility for coordination lies; and the overall nature of the building. Before embarking upon a discussion of these factors the prime objectives of the process of construction need to be stated. The process of construction has to ensure that the right materials, components, elements, labour, skills and aids to production (such as temporary works, plant and machinery) arrive on site in a sequence which will lead to efficient and safe working. To achieve this the process has to be preplanned and intentions communicated to all concerned. Various techniques are available, such as assembly drawings, written sequences, bar charts, flow charts, and critical path analysis. These give a visual display of the thought processes involved and are only produced as a result of analytical understanding of the construction process. This understanding relies upon the contributions of specialists, especially in complex buildings having a major input of service systems. The building technologist must have an overall understanding of the process, but will necessarily have to rely upon the specialists to provide information on the detailed sequencing. This knowledge needs to be known at the earliest planning stages so that any major implications arising from specialist inputs can be integrated into the programme, or the programme adjusted to suit.

As seen in the previous chapter, overall control of the process can be determined by the contractual arrangements adopted. For example, in the case of the traditional contractor/sub-contractor arrangement with supervising architect it is likely that the main contractor's programme was devised prior to contract, with little or no knowledge of the requirements of the sub-contractor. When the contract is awarded the main contractor then brings in the specialists and develops a detailed programme of work. It might transpire that assumptions made by the contractor were inaccurate and revisions to the programme will need to be made. If the services installation is complex and/or sensitive the whole construction programme could revolve around their sequencing. In this case the architect, at early design stage, should be aware of this and make the implications known to competing contractors: the responsibility for effective coordination of services during work on site rests initially with the architect.

In the case of project management the integration in process is achieved in a different manner. The responsibility for integration lies primarily with the project managers. They will determine the sequencing of work with a relatively clear understanding of the needs of the service engineers. The service engineers will be responsible for design and installation and will communicate their methods of work at tender stage, or before. Therefore, the coordination of their installation processes can be brought in at precontract stage. As overall control and responsibility is with the project managers they can ensure compliance with the overall programme. The contract (between specialist contractor and client) makes quite clear the obligations of the services contractor regarding the need to integrate fully with the process of construction. In assessing the tenders of services contractors the project managers will analyse the methods of working to ensure that they will fit into the programme, as well as checking for performance in function and adequate pricing. The services contractor's tender is evaluated on three counts: sequencing and methods of installation; matching functional performance requirements; giving value for money. Each of these will carry equal weighting. It may be that a number of contractors have produced tenders which are comparable on the latter two aspects, but that the distinguishing factor is the sequencing and methods of installation. In this case the project managers are likely to select the contractor who shows the most effective integration into the overall programme. There is an increasing tendency on the part of clients (and therefore of those acting on their behalf) to evaluate the on-site performance plans of all contractors in building. It is not acceptable just to be able to meet functional performance requirements at the right price but they must also meet the process requirements. Much construction work, whether new or refurbishment, has to be carried out within tight contract periods, and those contractors demonstrating that they can complete within time are more than likely to win the contract, even though their tender quotation may be higher than others. The client expects efficient building processes as well as an efficient building product (see British Property Federation 1983).

There are many potential problems to be avoided or reduced at the programming stage. These problems can arise due to inadequate forethought concerning sequencing operations or because of physical difficulties encountered whilst installing the services. These latter problems are discussed later; here the former will be analysed.

The sequencing of service installations is dependent upon a number of factors, as listed in Fig. 4.3.

Variety of service provision

If there are many types of service installation to be fitted into the building the task of integration can become extremely complex. What is to be fixed first? Does it depend on means of access, or method of fixing? What services can be run close together? If temporary works or other builder's work is required, when is it needed, and can it serve another service? Can continuity of attendance be given to each service installation? Does one service installation need to be complete before another can commence?

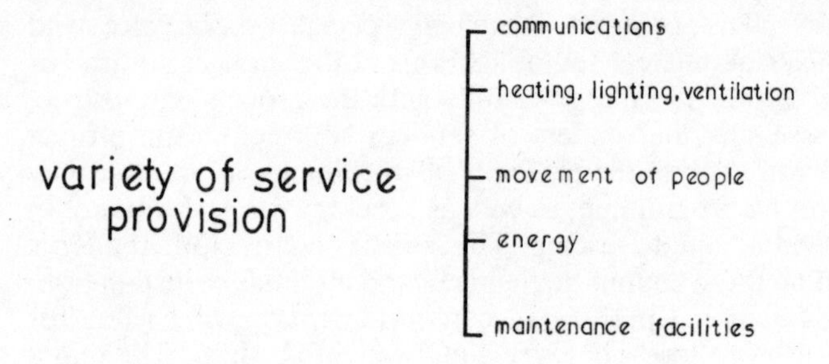

number of specialist contractors

contractual arrangements

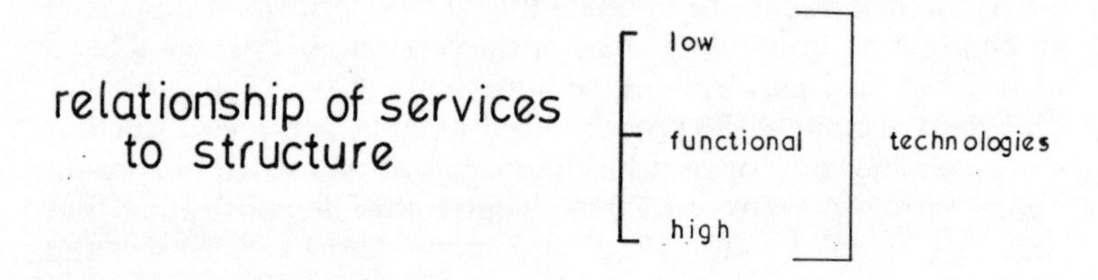

programming techniques

knowledge of methods

4.3 Factors determining service installation sequencing

Number of specialist contractors

The sheer number of specialist contractors can produce a complex integration matrix. Difficulties can arise through lack of effective communication between the contractors, and from the overall project coordinators. The prioritising of particular installations may be problematic: who goes first and when? How often and by whom should meetings be coordinated?

Contractual arrangements

This aspect has already been discussed in depth and the issues will need to be addressed in determining the service installation programmes. Depending on the degree and depth of involvement, the precontract programme will be influenced to a greater or less degree by the contractual arrangements.

Relationship of service installation to structure

If the services are a predominant feature of the building (high technology) then it is probable that they will set the pace in determining the sequence of operations. Where the building envelope provides most of the services functions (low technology) the sequencing is under the effective control of the building designer/producer. The lead is taken by the builder and the service engineers will need to fit into the overall programme. The functional technology approach demands an equal contribution from designer, builder and service installers. Each will have their preferences and the programme will need to be negotiated to suit their working practices and physical constraints.

Programming techniques

The manner in which the programme is prepared and presented can influence the validity of the final sequence. Bar charts show time and activity blocks with little or no indication of how they depend upon each other. They can be prepared using coarse estimates of time required. They may not show whether the activity is continuous or intermittent. If arrow diagrams or network analysis techniques are used a finer estimate of time is required, especially if being handled on a computer. These methods do allow a greater degree of detail and will show exactly which activity is dependent upon another finishing before it can commence. It must not be forgotten that the technique is only a visual representation of what should be expert knowledge of the processes involved and the time required to carry them out. The technology must be known before any programme can be presented.

Knowledge of processes

As mentioned above, a programme cannot be produced unless it is based on sound knowledge. At precontract stage the degree and depth of knowledge of service installation processes may be lacking, owing to: lack of hard information on appliances required; the programme being prepared by a person not familiar with all the processes; not all the specialist installers having been selected; the building structure design not having been finalised. Directly after the contract has been awarded all parties should be involved in preparing the programme. All service installers should be asked to provide detailed information on their sequencing and methods of work. The programme compiler should use this hard information and not rely on previous experience or assume that specialists can fit in to any time or sequence devised. In complex projects it may be necessary to produce a programme which indicates when and what type of information is required to achieve successful co-ordination of all the contractors, together with the designers. This programme sequences the process of co-ordination, not the process of physically constructing the building. The uses of this programme are: it will tell the designers when structural information is required; it will indicate to all parties the range of people involved; it will show what information is required before another activity can be determined; it will provide a basis for controlling and communicating the flow of information.

INTEGRATION WITH STRUCTURAL FORM

The basic structural form, such as frame, load-bearing walls, core structure, tension structure, or combinations of these, will influence the actual installation of the service provisions into the building. The floor and wall constructions will influence the methods of fixings and direction of pipes, etc. For example, a load-bearing wall may be restricted in the number of holes allowed through it, thereby constraining the run of pipes and cables. Timber joist floors should have the minimum of holes, with no notches either at the top or the bottom of the joists. Generally, framed buildings give a greater freedom for horizontal distribution ducts, but their form will be constrained by the types of floor/ceilings used. Is a false ceiling to be used to hide the services? How can the services be fixed to the floor/ceiling? When do these fixings have to be positioned, before or after formation of the floor/ceiling? Can fixings be made direct to the structural frame? How is the application of floor and wall finishes affected by the degree of completion of the service provisions?

Some simple illustrations will demonstrate the range and type of decisions that will need to be made when installing services. Electrical wiring to lighting and power points cannot be installed until the building is weathertight. But if it is a multi-storey building, can the installation be allowed to wait until the roof is completed? It might be better to introduce some temporary methods of waterproofing the lower floors, and to provide temporary external cladding to allow the electricians access. In this way the whole construction process can be speeded up, reducing the overall contract time and allowing the client earlier use of the building. Also, the contractors can get a quicker return on their resources and hopefully make a healthy profit. Electricity cables need to go to suitable power points etc. And these are positioned on firm backgrounds. Therefore, either the exact position of these backgrounds needs to be fixed and marked, or, preferably, the backgrounds should be in place. In addition, some chasing of grooves in walls might be required. The Building Regulations 1985 and Codes of Practice govern the extent and position of chases on structural walls, and note needs to be taken of these factors. If the cables are allowed to hang free prior to the backgrounds being placed, will they interfere with other activities? In some cases the cables need to be in position before the internal partition walls are completed as they will be built in during the erection of the walls. Before plastering brick, block or concrete walls, all services that are hidden need to be in place, together with their outlet boxes or junctions. After plastering, the outer covers or fittings can be placed, but should they be done before or after decoration?

Major items of service plant, or facilities such as escalators, may have to be placed before the envelope is completed because they require large openings for access. There may be other problems: for example, escalators are supplied and installed virtually self-finished. Care will need to be taken when handling and fixing the unit, and temporary protection provided to prevent damage by subsequent construction operations. This protection may in itself be extensive and will have to be programmed into the sequence of work, with implications for those activities in close proximity.

As the trend is towards more self-finished components and sophisticated service units in buildings (in preference to in situ fittings) the problem of their physical handling and integration into the structure will increase. Their use will depend upon the ease with which they can be installed. Large units requiring much space for access will need to be installed early; those items easily handled can be left until a late stage, thereby lessening the risk of damage. In order to handle some of the heavier units mechanical appliances may be needed: an escalator, to continue the example, has to be brought into the building on powered trolleys. Before that it will have to be lifted from its transport lorry; probably a mobile crane will be required for that operation. Can this crane swing the escalator into the building? What problems might this activity cause in the street? The timing of the operation may depend on external environment factors, such as an obligation to maintain access for the general public. Once inside the building and in close proximity to its working position, how is the escalator lifted into place? Can a small mobile crane gain access? Probably a powered block and tackle system will be used: to what can this be attached? Can it go into the building's structure or does it require temporary beams, etc? What energy source is required for the block and tackle? Will the main contractor need to be in attendance whilst this specialist installation is going on? If so, what skills will be required?

Some services will be built into the building's elements as the work proceeds. A common example is the placing of pipes and ducts through a solid ground floor slab. The services may also have to pass through the foundations to connect to the external mains. It is quite likely that the mains may not be in place at this stage and no connection can be made. In that case the two ends of the building's service need to be accurately located in space. Whose responsibility is this? That will depend upon the design, specification and contractual arrangement. Regardless of who does it, however, the task remains of ensuring that the pipes, ducts etc. are positioned correctly; this is where a structured set of drawings will help, as the 'location' series should give full dimensions, and references to datums and to the relevant 'assembly' drawings. The physical activity of placing the service pipes etc. is carried out simultaneously with the formation of the floor; this in itself can be a multi-stage activity involving sub-soil preparation, hardcore, blinding, damp-proof membrane, any insulation layer, and finally the concrete (which may have some steel reinforcement). The service pipes may have to be placed and temporarily supported prior to the work on the floor construction. Care needs to be taken to prevent displacement. Some service pipes and ducts will require a permanent protective surround: this will have to be placed prior to the floor slab. In some cases formwork boxes or polystyrene blocks will need to be placed to form holes, in order to house service items which cannot yet be positioned. Once a solid concrete ground floor slab has been placed it is extremely difficult to punch holes through it and burrow under the ground (and possibly through foundations) to provide a passage for a forgotten service. Suspended upper floors pose less of a problem for 'forgotten services', but ideally all holes should be formed at the time of forming the floor. If the floor is made up from precast units then the manufacturer should be informed of the positions of the holes and the units prefabricated with such accommodation provided. In situ concrete floors should have the holes formed at the time of placing by using boxes or polystyrene blocks, or by casting in the pipe or cable duct. When placing pipes and ducts in in situ concrete the open ends should have temporary stoppers to prevent the entry of concrete or debris. It is common to see pull-through cord in cable ducts; this should be securely tied at each end to stop it either being pulled out or slipping into the duct.

There are situations where service appliances penetrate external walls, for example, in the case of some types of heat exchangers. A major problem is that of ensuring the adequate weatherproofing of the appliance in the opening. Of course the design itself must be viable, but responsibility for the achievement of the integrity of the joint is shared between the builders of the wall and the service appliance installers. Both parties must ensure the opening is the right size and that the appliance fits as designed. But when should the appliances be fitted? Should they be installed in one operation, or can the external elements be fitted to provide weatherproofing and the internal parts left until the internal decoration stage?

The scale of distribution of the services can affect the way in which they are integrated with the structure, and this scale may in turn be influenced by the structural form.

What is meant by 'scale of distribution'? The term refers to the extent of the service runs and whether they are designed as one continuous system or are separated into self-contained systems. This concept can be applied to a single service, for example water; or to a facility, for example a toilet suite. Taking the case of water, the supply can be continuous pipe with short runs and taps off, and controlled from a single stop valve. Alternatively a main pipe can be run to each part of the building, and this is where the structural form can influence the layout. A compact building such as a two-storey house can be effectively served by one continuous water pipe run. A multi-storey building may have a main rising through a vertical duct, but with each floor having a separate sub-main. If it is further divided into, say, flats each may have its own run separately controlled. Buildings using a multiplicity of service appliances, such as hospitals or factories, may require separate water runs to each appliance. This will reduce the loss of service if an appliance fails; none of the others will be affected. Using a scale of distribution designed as separate sub-systems (whether in multi-storey or spread-out buildings) will allow an easier integration of the services, as they can be dealt with piecemeal. Each section can be installed, finished and tested independently of the others. The floors of a multi-storey building can be dealt with, if necessary, not in consecutive order, but in any order. In large area buildings sections can be isolated from each other and the services installed to match the progress of the overall construction activities.

Some buildings are designed so that service facilities such as toilets and washrooms may be installed as prefabricated, bolted-on, self-contained units. Under the concept of scale of distribution, these are considered small as they can be isolated from each other. Of course they need to be connected to the main service runs in the building, but each can be installed separately. Their position in the building's structure and the requirement that they be installed in relation to other building elements will influence their position in the construction sequence. If they are attached to the external face of the building then early installation may be necessary to ensure the watertightness of the envelope. Again, early installation may be required if they are positioned internally, as access may be difficult at a later stage in construction – the same problems may occur as were described in the example of the escalators. Accuracy and fitting tolerances are of paramount importance when installing prefabricated units, and many installation problems are generated by inaccurate sizing or inadequate (or mismatched) tolerances. This issue will now be discussed in as far as it relates to the installation of services in particular.

Five lessons have been drawn by Bonshor (*Jointing specification and achievement: a BRE Survey*, CP 28/77) from his work on obtaining suitable fit between components and elements. These are summarised as follows:

1. It is not acceptable to set only a single target value for joint widths.

Even with modified fixings it was found that uniform joints of specified width were achieved only by very close control of construction accuracy with, presumably, corresponding though hidden cost penalties.

2. Designs which identified and specified acceptable units of joint width did not usually attempt to identify corresponding inaccuracies which were tolerable for related stages of work.

There was little attempt to specify corresponding accuracies; too much reliance was placed on unrealistic expectations of accuracy; critical dimensions were identified but accuracy in them specified in an unworkable fashion; critical dimensions were wrongly identified; reliance was placed on 'designing out' the need for accuracy by avoiding 'contained' situations.

3. Tolerances are not usually thought of as a means of achieving satisfactory fit – except in the sense that vague statements are made about the need for perfect precision.

It was suggested that satisfactory fit is dependent on near, or even absolute precision, or that allowances will naturally occur.

4. 'Fixing adjustability' is a potential constraint on fit and should receive consideration at the design stages.

Fixings were poorly designed for adjustability. Where adjustability was adequate it was constrained by features on the components or even by the fixing itself.

5. Designers often believe that what is specified or shown on a drawing can and will be achieved on site. Builders often hold a different but equally simple view, that anything fits which can be put – or forced – into place.

Designers should be expected to specify acceptable limits. Without such specification, builders cannot know what designerss consider as satisfactory fit. Designers usually only think in terms of an 'average' joint width – taken to be the dimension on the drawing. In practice the chance that this value will prove to be the mean – still less the commonest width achieved – is very remote.

Obviously there are some major problems regarding the achievement of satisfactory joints and it is to be hoped that since this survey was published in 1977 improvements have occurred. Bonshor concludes that 'design for fit is not at present based on any sound rationale. Drawings are confused with reality, with designers expecting that notional modular lines on a drawing have an actual existence on site and that if work is built exactly as drawn, fit is assured. If joints in buildings are to function satisfactorily, designers must understand the factors affecting fit and be able to manipulate them so as to improve the likelihood that achievement will match specification.'

The BRE has published a handbook to aid designers in creating acceptable tolerances and fits (Bonshor, R.B. and Eldridge, L.L., 1974a). This shows how best, by using diagrams, graphs and illustrative sketches, to design joints for general component and material assemblies. Supplementing this handbook is a BRE Current Paper (Bonshor and Eldridge, 1974b) reviewing the problems which may be met in any application of the principles. A number of common misconceptions are disposed of in these two publications, as follows.

- there is no need to design explicitly for fit;
- there is certainly no need for the system to be used in traditional construction: it is only advantageous where most or all of the components are prefabricated, especially when those components are also standard, and bought in the open market;
- there is no need to specify tight tolerances. Specification needs only to be close enough for the finished work to function adequately;
- components delivered on site must be considered as typical of the sample. Even if the sample is of a single component, as it comes from the general component population it must be deemed to be representative;
- further misconceptions are dispelled in the Current Paper, which covers details in tables and the proving of the aids in practice, and includes a discussion on the issue of tolerance as a synonym for inaccuracy.

The conclusions arising from these studies by BRE imply a lack of knowledge on the part of designers, with a simplistic view of jointing being taken by builders and installers of components. In the first place the design must be practical and the drawings prepared must give the appropriate dimensional information. If these are produced in accordance with the BRE handbook the joints, filling and tolerances should end up within feasible and practical ranges. In order for the builder and installers of service items to ensure adequate fit they should therefore be familiar with the principles stated in the handbook. Good fit and functional efficiency can only be achieved if both designer and installer/assembler are cognisant of the problems and possibilities. It is probable that many designers will be involved in producing service appliances and ancillary distribution networks. These will also have to be related to the structure.

Generally, service appliances are manufactured to a higher level of dimensional precision than say, in situ concrete work. What is feasible for metal is not feasible for site concrete. When actually preparing holes, placing fixings etc. for service appliances the builder must ensure their accuracy by:

- checking drawings (from architect, appliance manufacturer, services designer, structural engineer, temporary works coordinator) for viability of dimensions (against BRE handbook) and relationship to other services;
- ascertaining sequence of installation;
- setting out accurately on site;
- checking setting out against drawings and on-site positions;
- forming holes, fixing etc. with specified fittings, linings etc.;
- assuring that appliances etc. have been manufactured to prescribed tolerances;
- providing competent supervision of builder's work in connection with services installations;
- checking that any instruments used in setting out are calibrated correctly and are used in accordance with standard recommended procedures (see Dean, W.B. and Stevens, A.J., *Accuracy achieved in setting out with the theodolite and surveyor's level on building sites*, BRE Current Paper 15/77, for observations on the levels of accuracy obtained);
- ensuring that seals, joints, covers are positioned correctly and made with the right materials;
- ensuring adequate and effective testing and commissioning procedures are carried out.

This last item will now be considered in greater detail.

COMMISSIONING AND ACCEPTANCE TESTING

The bringing into use of the services of a building is a two-stage process, commencing with commissioning and followed by acceptance testing. Commissioning is the process of starting up the services from a unit position. Where appropriate the items, appliances and service output are regulated to produce their designed performance. Acceptance testing is the process of evaluating their performance. How effective are they? Do they produce the required environment? Do they meet the user requirements?

There are a number of issues associated with the processes of commissioning and testing. Their implementation is essential to ensure a proper handover of the service installations to the client or building users. The elements of these processes are now described with a brief comment on each.

Commissioning tests

These are essentially the responsibility of the contractor (or installer) of the system. Overall responsibility lies with the main contractor, and he might well need to coordinate a number of subcontractors to ensure there are no conflicts in the commissioning procedures. The nature of the commissioning and the programme has to be agreed by all parties.

Acceptance tests

The type and integrity of the tests should be laid down by the services designers, bearing in mind the user's performance criteria. The tests should be clearly described, executed by the contractor and witnessed by others.

Commissioning team

Specialist, independent teams can be employed to carry out the commissioning and testing stages. They will act on behalf of the client and be directly responsible to him. These teams may consist of specialist engineers, with varied expertise reflecting the different service installations.

Manufacturers' representatives

It is common practice for the appliance (or system) manufacturers to be in attendance during the commissioning and testing stages. They might even initiate some of the procedures. These representatives must be experienced in these processes so that they can cope with any problems.

Materials and power

Responsibility for the provision of any materials and power for commissioning and testing should be clearly defined, together with liability for the costs. Such materials as gas, water and fuels should be available, together with electricity. It may be necessary to provide some materials and power on a temporary basis, such as electricity from portable generators.

Safety

A major issue relating to services is the safety factor. Here the concern is at two stages, during the commissioning and testing procedures, and during the initial performance of the installations. During the testing stage the following three rules should be observed: one, only experienced staff should carry out testing; two, the commencement and duration of the testing procedures should be communicated to all on site (it might involve the closure of all or part of the building to other trades); three, in the case of failure, emergency procedures should be rehearsed. There are other issues connected with safety, such as: compliance with statutory legislation; fire protection and control; provision of warning notices; security of installations; safety of temporary services; and the prevailing weather conditions can produce adverse affects if the services are tested 'out of season'.

Manufacturers' warranties

If a manufacturer has given a warranty its terms should be upheld in case anything should go wrong. The builder/specialist contractor should ensure that the warranty is not invalidated by undertaking modifications and/or maintenance not in accordance with the manufacturer's instructions. Initially the warranty will have been given to the builder/installer, as they paid for the appliance, so it might need to be assigned to the client or building's user.

Spare parts

The manufacturers/suppliers should keep a stock of spare parts of those items likely to wear or fail. This stock should be available for at least two years after the defects liability period. A comprehensive list should be provided.

Operating staff

Ideally the operating staff should be in attendance during the commissioning and testing stages, but this is not always possible. Adequate training programmes should be devised, and for large complex systems the manufacturer/installer may well provide people for a limited period of time to operate the plant and train the permanent operating staff.

Forms and records

A complete record of the tests carried out during the commissioning and testing stages must be kept. The use of standard forms can aid the correct procedure for tests; these can be used as a basis for future tests. The commissioning and testing stages are an integral part of the installation process and are covered by the contract conditions. Consequently, the forms and records will comprise a part of the contract documentation. These should be held on file where they can be easily retrieved by the users of the building.

Manuals

Operating manuals may be supplied by the manufacturers/installers for each appliance or system. Although these will cover the individual system they are rarely produced in the context of the whole building functioning as an entity. Little reference is made to such aspects as the opening of doors (its effect on heating) or the effect of uncommon activities by users. The provision of manuals for purchasers of houses is advocated by consumer associations. These will give the owners instructions on operating the services, such as central heating, and will also give guidance on the maintenance of the performance of the whole house, including care of the fabric.

QUALITY ASSURANCE

Quality assurance is the process whereby the highest levels of integrity are awarded to an item or complete system. A rigorous procedure of checks and rechecks has to be carried out to ensure that the required standards have been achieved. The process is being adopted by some house builders as a guarantee that their products will perform to the desired standards, and will not generally fail – but that if they do, remedial work will be carried out at the builder's expense. The cost of providing an effective quality assurance process can add up to 30% to the normal costs, but in some cases this can be justified, as a system failure could be disastrous to the building's owners and users.

Commissioning and testing is a vital part of the installation process for service systems. It needs to be effectively integrated into the construction programme and could entail a lengthy period of time. Some of the tests could be carried out immediately after installation but the majority will have to take place during the later stages of the work, as the various systems will need to be balanced and set to work as a whole within the total building structure.

SUMMARY

Integration with construction is a two-part activity. One part is concerned with physically fitting the service components into the building, the other with fitting the installations into the planning process of the work. Constraints operate within both activities and the builder must be aware of these and organise the work so as to limit their effects. Although integration is here analysed in two parts, in reality the two must be considered as a whole. Problems in placing the fixings are largely a matter of when they are placed, how, and by whom. We considered the example of pipe fixings to the underside of a concrete suspended floor. Are they to be placed during the floor construction? How accurate do they need to be? Who will place them? Who takes responsibility for accuracy? Can they be placed after completion of the floor? What form of attachment can be used for the fixing to the floor? Will the reinforcement in the floor (and other buried items) restrict the positioning of the fixings? If the fixings are placed later, rather than during floor construction, what will be the effect on the sequencing of other trades? What access and power facilities will be required? The foregoing questions illustrate the interrelationship in the course of installation between physical and procedural aspects. Finally, an essential part of the installation process is commissioning and acceptance testing to the user's requirements, and this brings together the building's elements of structure and fabric and the service systems.

QUESTIONS

1. Discuss the major differences between installing services in an existing building and in a new building.
2. Describe the responsibilities of a site manager with respect to the installation of services by specialist sub-contractors.

Part Five
CASE STUDIES

15. Three Construction Case Studies

INTRODUCTION

The purpose of the following case studies is to show in practice some of the concepts, issues and technologies described in the text. The actual solutions adopted by the builders, designers and consultants are given; these studies are based on real construction projects. The solutions used and described were developed using the knowledge and technologies available at the time, but it is likely that alternatives are possible. The sequence of operations could be changed to ease a particular problem. These studies should not be read as the only solution, but should be continuously questioned as to their methods and related to your own experience and knowledge. From time to time in the descriptions questions will be posed to stimulate the development of alternative strategies. After you have read and analysed the studies a completely different solution could well present itself, especially in the wake of technological developments which may have arisen since the writing of this book.

CONSTRUCTION STUDY ONE – 70 FLATS, GARAGES AND A BANK

This study is written from the perspective of aids to production. The main methods, plant, machinery and sequences are described for the construction of the flats, integral bank and garages, with mention of the demolition of the existing cinema and bank on the site (Fig. 5.1).

Structures

The 70 flats are in two blocks, one three-storey, the other two-storey, to take account of the step in the site level. The existing levels were maintained to reduce cut and fill, and the roof levels of the two blocks are therefore approximately level. The walls are all of load-bearing brickwork, with precast concrete wide-beam suspended floors and in situ concrete communal stairs and landings. The ground floor is solid concrete. The roof is flat constructed with timber joists, fixings, pre-felted wood wool slabs and two layers of felt. The walls are plaster finished with internal partitions. Some garages were built integral at ground floor of the three-storey block. The other garages were constructed of in situ concrete columns and beams carrying precast concrete beams for additional car parking over. The new bank, integral with the three-storey block, was formed of in situ concrete columns and beams, with in situ concrete floor over supporting the flats above. Cladding to the bank was in precast concrete panels. Two bridges allow access to the upper two floors of the three-storey block, together with a staircase. All were constructed in in situ concrete with brick balustrading. Foundations to the flats were concrete strip with brickwork up to d.p.c. The columns had independent concrete bases.

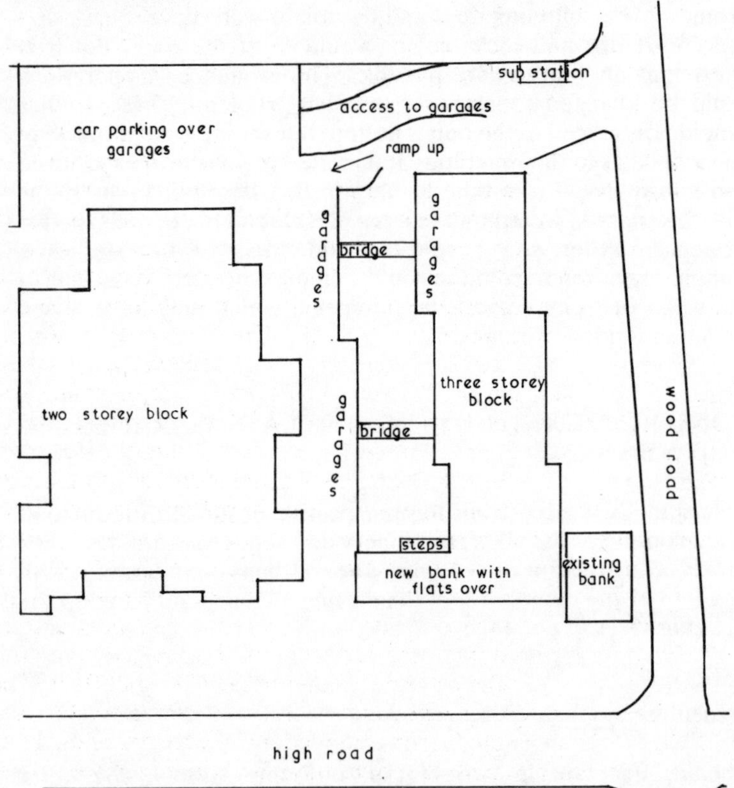

5.1 Construction Study One: flats, bank and garages

Site

The site was naturally split into two levels. On the lower level there was a cinema with integral bank. At one time, on the upper level, houses stood fronting on to High Road; these had been demolished prior to the client acquiring the site. Access to the site, both for temporary and permanent use, is from Wood Road, adjacent to the already existing sub-station.

Description of technologies

1. Demolition
This work was carried out by a specialist sub-contractor nominated by the architect. The primary method of demolition was by a ball and chain suspended from a mobile crane jib. Not all the building could be demolished as the integral bank had to remain in use until the new structure was complete. As demolition got nearer to this part of the building hand work was necessary both to ensure collapse did not occur, and to reduce noise for those people working in the bank. As internal walls were exposed by the demolition of the cinema section they had to be given protection from the effects of the weather; a coat of bituminous paint was applied. The bank's vaults extended under the cinema and into the site for the new bank, so they could no longer be used. A temporary single-storey brick cavity wall extension was built on to the site of the existing bank to provide vault accommodation; this work was carried out by the demolition contractor. The demolition was taken down to ground level, leaving the ground floor slab and foundations to be taken out by the main contractor during excavation for the new foundations. Two problems arose which had to be dealt with by the main contractor, one of which was unforeseen. The known problem was that of the old vaults, which were directly under the position of two new columns. The vault walls were dug out, care being taken not to disturb the foundations under the existing bank; this left a hole approximately 3m deep. The normal depth of foundation pad to the columns was 1m, which meant that 2m of depth needed to be made up. The solution adopted was to construct a box in ply and timber formwork to the dimensions of the independent pad foundations. A weak mix of concrete was poured up to the underside of the designed pad, this concrete resting on the exposed ground below the vault floor. The reinforcement and designed concrete mix was then placed on top of the weak concrete, again retained by formwork already in place. Alternative methods could have been used: for example, filling the complete hole left by the vault with well-consolidated fill to the underside of the new pad foundations, or placing a weak

concrete mix in the complete hole. What factors would influence the method of constructing the new foundations?

The unforeseen problem was discovered when the cinema ground floor slab was being taken up. A hole appeared which was found to be built-in brick and concrete walls and floor housing an old heating boiler and subsidiary pipes and fittings. This hole was situated squarely under a large area of the new three-storey block. It appeared that a newer heating system had been installed in the cinema and that the old system had not been removed, access to it being sealed with a concrete slab. The solution was to take out all the equipment and the structure, leaving a large deep hole in the middle of the new foundations. The consultant structural engineer produced a new design for the foundations in this area. In essence this was an adaptation of the original design, that is, concrete strip with brickwork up to d.p.c. In this case the concrete strips were placed in the bottom of the hole (about 2m deep) and stepped to join the normal strip at 800mm below ground level. The brickwork up to d.p.c. was in engineering bricks; the external walls one and a half bricks thick; the internal cross walls one brick thick. (The external walls above d.p.c. reverted to the normal two-leaf and cavity construction.) The space within the walls was filled with clean hardcore placed in well consolidated layers. The ground floor was designed as a suspended concrete slab, so reinforcement was placed in it. This measure was undertaken to mitigate the effects of the fill sinking.

Was it necessary to use engineering bricks up to d.p.c.? Could the whole area of the hole have been filled with weak concrete up to the underside of the normal foundations? Would it have been better to use precast concrete beams for the floor?

This section has illustrated the following: contractual responsibility; demolition techniques; problems arising from existing buildings and their effect on new building work. See Fig. 5.2 for details of the technologies employed.

2. Garages and bridges

The garages under the three-storey block are integral with the load-bearing brick walls and do not cause any particular difficulties. Those opposite and to the top left-hand corner of the site do pose some problems with regard to aids to production. The large set of garages at the top of the set was excavated out of the higher level ground: they required retaining walls to all four sides with access through one, adjacent to the sub-station. These were formed in concrete with traditional ply and timber formwork. The curved retaining wall supporting the ramp up to parking area on the roof of the garages diminished in height as it followed the slope of the ramp and also tapered in its thickness from base to top. Again, traditional ply and timber formwork was used, but constructed in straight sections so that the finished wall was not truly curved but was a series of straight sections. These features were disguised by building a facing brick skin on the side exposed to the access road to the garages. The main reason for constructing the formwork this way was economy. To form the ply into curves would have been difficult and costly. Would sheet metal have been a cheaper and/or easier material to use to create the required curve?

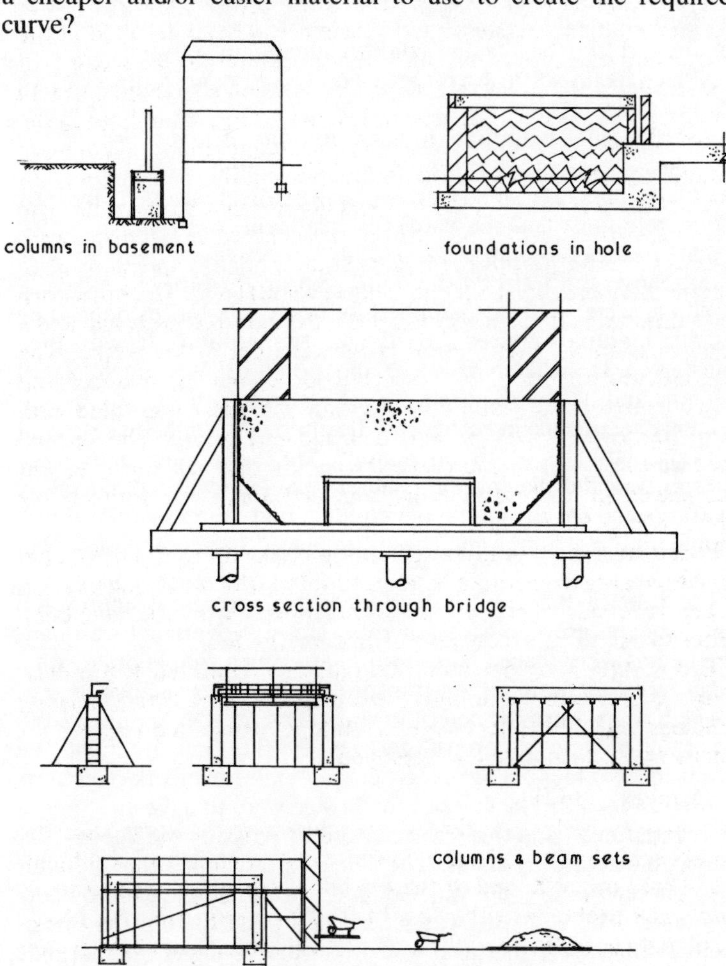

5.2 Construction Study One: technologies employed

The process for forming the columns and beams was as follows.

1. Excavate independent column bases by hand.
2. Place reinforcement and pour concrete.
3. Form kicker on concrete pad.

4. Fix column steel work to starter bars.
5. Place ply and timber formwork for column sides, secure with adjustable cramps and plumb and prop with adjustable props.
6. Erect platform for access and fix ladder.
7. Pour concrete via buckets and vibrate.
8. Leave columns to cure, then strip formwork.
9. Erect working platform between columns for access to forming beams.
10. Fix steel for beams and connect to column starters.
11. Place beam soffit, form, level, on bearers and props.
12. Fix beam sides.
13. Secure sides with bolts to each other at correct width.
14. Check level and depth of concrete required.
15. Place concrete and vibrate, using wheelbarrows lifted by mobile hoist and run via access platforms to beam.
16. Strip sides of beams.
17. Remove soffit after 7 days and re-prop for further 21 days.

All concrete used was supplied by ready-mix lorries depositing the load onto prepared boards on the ground. It was then transferred by wheelbarrow to the point of use. Each set of columns and beams required $2m^3$ of concrete.

Can the sequence in such a case be altered? Can the necessity of working platforms be circumvented? Are there other methods of transporting and placing the concrete economically? What safety measures and checks should be implemented for these operations? Should all the columns and beams be placed simultaneously or set by set?

The roof of the garages was formed of precast concrete and hollow beams grouted with concrete. This was overlaid by asphalt. The beams were placed by the supplier using a simple pulley on a tower to lift them, and then transported on a two-wheeled bogey.

The bridges were formed of in situ concrete; they spanned from the roof of one group of garages to the first floor of the three-storey block. The cross-section and formwork layout is shown in Fig. 5.2. The sides of the bridge were to have an exposed aggregate finish, and this was achieved by painting the internal ply face with a retarding agent. The sides were stripped the day after the concrete pour and washed and wire-brushed to show the aggregate. The concrete was placed and poured from the bridge itself. Is this a safe method? What alternative might be adopted? What sort of falsework should be used to support the bridge formwork?

General site technologies

Putlog scaffolding was used around the brick walls for the

construction of the two- and three-storey blocks. All the mortar was mixed on site, the cement being delivered in bags and sand placed adjacent to the mixer. An adequate store shed was provided for the cement. The usual welfare facilities were provided.

Supplementary questions

1. The floor above the new bank was designed by the structural engineer to be in situ concrete as it had to transfer the loads of the flats above to the columns forming the bank. The external edge of this floor was of a similar profile and finish to the bridges previously described. Furthermore it was to be placed in one pour as no construction joints were allowed; the total volume of concrete was $95m^3$. Describe what would be a suitable falsework to support the floor and its integral beams. Produce a strategy to ensure this falsework is adequate to meet the demands placed upon it, including the pour, curing and stripping operations.
2. Considering the site as a whole, produce a sequence of operations which will utilise efficiently such aids to production as scaffolding and formwork, and mechanical excavation plant. For example, should the excavation to the large garage area be carried out simultaneously with the foundations to the blocks? Should both blocks' superstructure be commenced simultaneously?

CONSTRUCTION STUDY TWO – OFFICE REFURBISHMENT

The refurbishment of this city centre office block was planned around the new service installations. The building's stone-clad neoclassical facade was unaffected, all work taking place within the seven-storey structure while a number of tenants remained. These tenants, and those expected to take up the remaining space, were financial institutions reliant on sophisticated communications networks and needing the usual toilet facilities. Also, in keeping with the demand for a completely controlled internal environment, an air conditioning system was to be installed. There was a central lightwell to the building and permission was granted to fill this in with floors at every level to provide more lettable office space and plant accommodation. The main plant room was situated in the basement. At 1987 prices the value of the contract was £7 million. The scope of the services is indicated by the fact that nearly 600 drawings were produced. The work was undertaken to a standard form of contract under the supervision of a construction management firm. There was a main building contractor experienced in this type of work, with a major sub-contractor for the services installation. Consultants were employed for the building services, construction costing and structural engineering.

Structure

The main structural work was the filling in of the existing lightwell at each floor level to existing roof level. A completely new structural steel frame was erected in the well, supported on new concrete strip foundations. This was designed to support the new floors and give support to the existing floors. This was necessary because the walls to the lightwell were load-bearing. In order to allow open internal space these load-bearing walls were demolished, and the ends of the floors stiffened and then supported on the new structural steel frame. In order to carry this out safely a complex, specially designed temporary support system in proprietary steel sections was used. All floors were propped simultaneously, the vertical props being supported (with load transference) by steel needle beams. As tenants were still occupying part of the building, temporary walls were required to the lightwell prior to its being roofed and floored. Timber stud partitions, each side faced in plywood and insulated with rockwool, provided adequate protection. Additionally, escape routes had to be maintained throughout the refurbishment and these had to go through areas where work was taking place. Again, stud partitions were used to delineate the escape corridors. Fig. 5.3 gives an outline of this work.

Services installations

An added complication in the upgrading and provision of new services was the fact that some tenants still occupied the building during the work, and of course, their service needs had to be met. The building had an old air conditioning system and this was to be replaced. The chillers to the new system were ultimately to be placed on the roof, but to provide a current service the chillers were placed on a temporary cantilevered platform over the street at fourth-floor level.

An extensive pre-contract survey was carried out to ascertain the position and routing of the existing services, but despite this many were found to be ‘popping up all over the place’. Much work

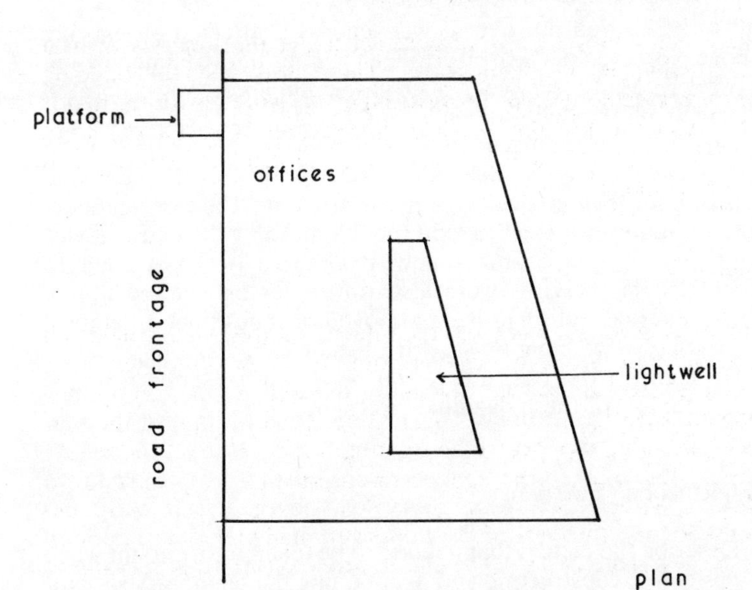

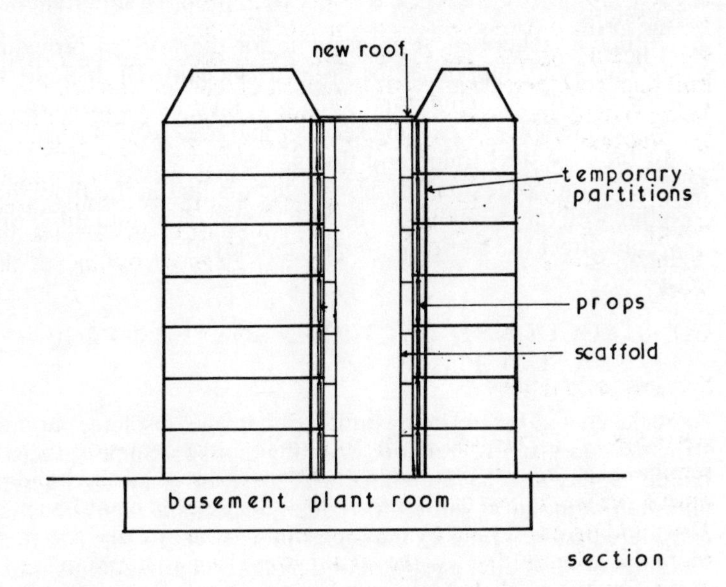

5.3 Construction Study Two: office refurbishment

had to be carried out over a weekend as it affected an existing service, for example work to toilet facilities and communication networks. Many services, such as drainpipes, were located externally on the lightwell walls, and these had to be maintained. Special fixings were placed on the temporary timber stud walls to hold the services. This was another reason for providing the stud walls and for them to be of strong construction. The existing floors were of hollow pot construction, unable to carry the extra service equipment loadings, and a new steel grillage was needed to support them. Services were also run under new raised floors. Where internal walls had been removed or repositioned, temporary services were hung on steel trestles or hangers.

The contract was over a two-year period with a series of staged completions. All these deadlines were met.

Supplementary questions

1. Describe the factors that needed to be taken into account when designing, constructing and maintaining the temporary support platform for the air conditioning chillers, with special reference to safety.
2. Outline a technological programme for the work of providing the temporary support to the floors in the lightwell. Include the provision of the temporary walls and take into consideration the maintenance of existing services and the erection processes for the new steel frame and floors.
3. Discuss the contextual framework factors that in the first place influenced the decision to refurbish the building, and those that would affect the site technologies employed.

CONSTRUCTION STUDY THREE – CONVERSION OF WAREHOUSE TO FLATS

As social change occurs many buildings become obsolete, but their structure and fabric remain basically sound. Some of these buildings may also have architectural merit or historical interest, and their demolition will be forbidden, especially where they are considered to be prime examples of their type. In the UK these buildings are listed on a register and assigned a grade. Only minor alterations may be undertaken to a listed building and the main structure, together with its components and finishes, must be maintained in its original state. The town centre warehouse in this study was just such a building. The town is a seaport which no longer has any trade requiring warehouses, especially those built in the late 1800s. Consequently, an alternative use had to be found for this one.

The builder was already the owner of the warehouse, and proposed a scheme to convert it to 28 flats. As the building was seen to be of merit it was hoped that a number of grants could be obtained to help with the costs of conversion; the remainder of the money was raised privately. All the flats were to be sold, and the majority were presold. Two grants were obtained, one an Urban Development Grant from the government, comprising 30% of the costs, the other from English Heritage to help in restoring the façade, amounting to 2% of the total costs. The architect was directly employed by the client/builder.

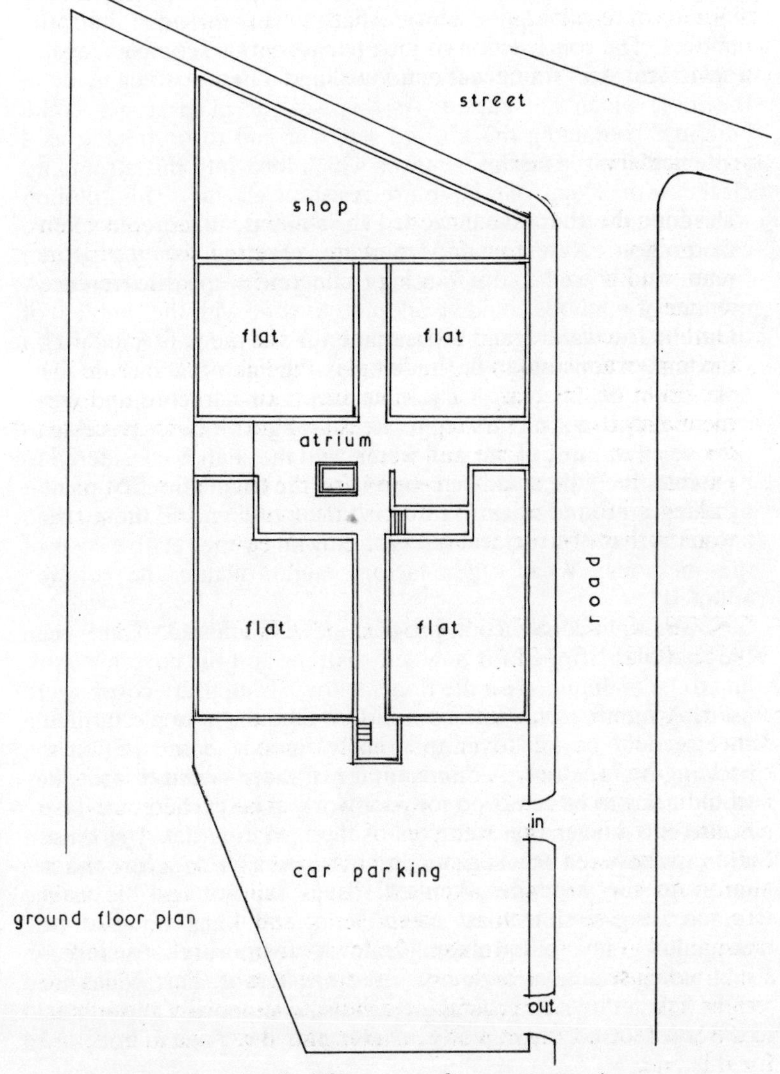

5.4 Construction Study Three: warehouse conversion

Structural work

Fig. 5.4 shows the floor plan of the converted four-storey brick warehouse. A shop and car parking was situated at ground level. An atrium was created in the middle of the building, effectively splitting the building in two. This also created an architectural feature as two sides were glazed, giving views across the town, and a welcoming entrance court. The main vertical services were carried up here, and were distributed horizontally along the corridors (in ceiling voids) to the flat front doors. Many of the floors and walls were unable to carry the new loadings and a large amount of structural steelwork had to be threaded into the building. The construction of the atrium was also complex; again, a structural steel frame was employed and a new opening made in the roof, which was glazed. As this was the main access to the building, containing the lift and stairs, it had to be treated as a protected shaft. The flat windows which look into this atrium are glazed with 20mm one-hour fire-resistant glazing. This solution was adopted rather than enclosing the staircase in a fire-protective construction. New foundations were required for the atrium frame, and a well and a block of concrete were discovered. A number of methods could be adopted to cope with the problem of the well. The water could be pumped out and the hole filled with a weak mix of concrete to the underside of the new foundations. The hole could be filled with alternate layers of hardcore and weak concrete, with a structural concrete capping. If it is not possible to remove all or most of the well water, and the shaft is considered to be stable, the hole could be capped and the foundations oversail it by using a ground beam. A major factor influencing the method chosen is that of structural safety, followed by the relative costs of the methods. What other factors might dictate the solution adopted?

A very tight construction programme was adhered to and when the roof was stripped for new felt, battens and tile covering work had to be maintained on the floors below. Temporary covers were used. A number of solutions could be adopted. Simple tarpaulin sheets could be laid over the roof rafters, and secured to the building by rope ties. A disadvantage of these would be that they would need to be removed for roof work to be carried out. There is also the danger that rain could seep in during heavy storms at overlaps between sheets. It might not be possible to secure the ties safety to the structure. Applied fixings are not feasible as the façade is of architectural merit, so the holding down of the tarpaulins may be problematical. A temporary structure in scaffolding could be erected over the whole roof. This would need to be taken down to ground level for safe support. The top could be roofed in corrugated steel sheets and the sides at roof level enclosed with polythene sheeting. The support scaffolding could also be used for work to the façade, although this may cause problems at ground level, as the building fronts onto street pavements on three sides. Should protected walkways be provided, or the pavement area sealed off and an alternative walkway provided? What factors would govern the decision? A third method of providing weather protection at roof level would be to construct a false flat roof within the existing pitched roof. Simple scaffolding framework techniques could be used, overlaid with polythene or tarpaulins. Disadvantages here are the need to drain the water off this false roof, and to close this area off to any other work, such as the installation of pipes and tanks. Depending on the strength of the existing structure it might be possible to erect a lightweight, partially cantilevered framework at roof level to give a structure for a false roof, but this option is not feasible here as the warehouse structure is relatively weak; hence the need for the steelwork inserts already mentioned.

The site can be classed as restricted, all materials having to come into the car parking yard. Deliveries would need to be carefully controlled as storage space is at a premium. What arrangements can be made for the site personnel accommodation, such as offices and health and welfare facilities?

Supplementary questions

1. Outline the sequence of work for the following major work packages, giving the aids for production necessary for each: atrium; restoration of the façade; strengthening of the floors; erection of partitions forming flats; first fix services; second fix services; finishes; roof re-covering.
2. List and describe the site accommodation for the project, based on the assumption that the maximum number of operatives at any one time will be 32, including three office-based staff.
3. The services for this project amounted to 22% of the contract sum, a relatively low percentage when compared to like conversions. The heating was supplied by electric storage heaters and hot water via electric immersion heaters. A lift was provided. Discuss the role of the services for this project and speculate how these may be enhanced to give a high standard and range of provision. Consider such aspects as security, waste disposal and internal environmental control.

References and Further Reading

Books and articles

AMBROSE, J., *Building Structures Primer*, John Wiley and Sons, 1981.

ARMSTRONG, B., *Programming of Building Contracts*, Northwood Publications Limited, 1981.

ATKIN, B., 'Potential of Robotics in Design and Construction', *Chartered Quantity Surveyor* (9) January 1987, p. 9.

ATKINSON, G., 'Designed by Humans – Built by Robots', *Building* (252) February 20th 1987, pp 50–51.

BARLAND, J.B. and HANCOCK, R.J.R., *Underground Car Park at the House of Commons, London: Geotechnical Aspects*, British Research Establishment Current Paper 13/77.

BENTLEY, M.J.C., *Quality Control on Building Sites*, Building Research Establishment Current Paper 7/81.

BONSHOR, R.B., *Jointing Specification and Achievement: A British Research Establishment Survey*, Building Research Establishment Current Paper 28/77.

BONSHOR, R.B. and ELDRIDGE, L.L., *Graphical Aids for Tolerances and Fits: Handbook for Manufacturers, Designers and Builders*, Building Research Establishment, HMSO, 1974a.

BONSHOR, R.B. and ELDRIDGE, L.L., *Tolerances and fits for standard building components*, Building Research Establishment Current Paper 65/74, 1974b.

BOWYER, J.T., *Small Works Supervision*, The Architectural Press Limited, 1975.

BOWYER, J.T., *History of Building*, Orion Books, 1983.

BOWLEY, M., *The British Building Industry: Four Case Studies*, Cambridge University Press, 1966.

BRAGG, I.L., *Final Report of Advisory Committee on Falsework*, Health and Safety Executive, June 1976, HMSO.

BRAND, R.E., *Falsework and Access Scaffolds in Tubular Steel*, McGraw Hill, 1975.

BRITISH PROPERTY FEDERATION, *Manual of the British Property Federation System: The British Property Federation System for Building Design and Construction*, British Property Federation, 1983.

BROWN, M.A., *Demolition*, CIOB Occasional Paper No. 17, Chartered Institute of Building, 1978.

BRANDON, P.S. and POWELL, J.A., *Quality and Profit in Building Design*, E. and F.N. Spon, 1984.

BRE, *Energy Consumption and Conservation in Buildings*, Digest 191, July 1976, Building Research Establishment.

BRE, *Getting Good Fit*, Digest 199, March 1977, Building Research Establishment.

BRE, *Accuracy in Setting Out*, Digest 234, February 1980, Building Research Establishment.

BRE, *Waste of Building Materials*, Digest 247, March 1981, Building Research Establishment.

BRE, *Materials Control to Avoid Waste*, Digest 259, March 1982, Building Research Establishment.

BRE, *Common Defects in Low-Rise Traditional Housing*, Digest 268, December 1982, Building Research Establishment.

BRE, *Fill. Part 2. Site Investigation, Ground Improvement and Foundation Design*, Digest 275, July 1983, Building Research Establishment.

BRE, *Site Investigation for Low-Rise Building: Desk Studies*, Digest 318, February 1987, Building Research Establishment.

BRE, *Security*, Digest 122, October 1970, Building Research Establishment.

BRE, *Principles of Joint Design*, Digest 137, 1977, Building Research Establishment.

BRE, *Electricity Distribution On Sites*, Digest 179, July 1975, Building Research Establishment.

BRE, *Demolition and Construction Noise*, Digest 184, December 1975, Building Research Establishment.

BRE, *Working Drawings*, Digest 172, December 1974, Building Research Establishment.

BSI, *Construction Drawing Practice*, BS 1192 Parts 1–4, 1984, British Standards Institution.

BSI, *Code of Practice: Safety in Erecting Structural Frames*, BS 5531, 1978, British Standards Institution.

BSI, *Guide to British Standard Codes of Practice for Building Services*, BS 5997, 1980, British Standards Institution.

BSI, *Demolition*, Code of Practice CP94, 1971, British Standards Institution.

CHANDLER, I.E., *Materials Management on Construction Sites*, The Construction Press, 1978.

CIRIA, *A Suggested Design Procedure for Accuracy in Building*, Construction Industry Research and Information Association Technical Note 113, 1983.

COLLIER, K., *Managing Construction Contracts*, Reston Publishing Company Inc., 1982.

COVINGTON, S.A., *The Degree of Quality Assurance Provided with Certain Building Components and Products*, Building Research Establishment Current Paper 8/80.

COURTNEY, R.G. and HOBSON, P.J., *The Performance of 15 District Heating Schemes*, Building Research Establishment Current Paper 13/77.

CRAWSHAW, D.T., *Co-ordinating Working Drawings*, Building Research Establishment Current Paper 60/76.

CUTLER, L.C. and S.S., *Handbook of Housing Systems for Designers and Developers*, Van Nostrand Reinhold Company, 1974.

DALTRY, C.P. and CRAWSHAW, D.T., *Working Drawings in Use*, Building Research Establishment Current Paper 18/73.

DEAN, W.B. and STEVENS, A.J., *Accuracy Achieved in Setting Out with the Theodolite and Surveyor's Level on Building Sites*, Building Research Establishment Current Paper 15/77.

EDMEADES, D.A., *The Construction Site*, The Estates Gazette, 1972.

ERSKINE-MURRAY, P.E., *Construction Planning – Mainly a Question of How*, CIOB Occasional Paper No. 2, 1976, Chartered Institute of Building.

ESHER, L., *The Broken Wave*, Penguin Books, 1981.

ESHETE, S. and LANGFORD, D., 'Planning Techniques for Construction Projects', *Building Technology & Management* (25) March 1987, pp. 30–31.

FERGUSON, I., 'Buildability. The Influence of Design Upon Building Method 2. Case Study: A Steel Framed Building', *Building Technology and Management* (25) March 1987, pp. 32–34.

GARDNER, E.M. and SMITH, M.A., *Energy Costs of House Construction*, Building Research Establishment Current Paper 47/76.

GIBSON, D.H. and HIGGS, T.A., *The Building Regulations 1985 – An Introduction*, CIOB Technical Information Service No. 76, 1987, Chartered Institute of Building.

GOOD, K.R., *Handling Materials on Site*, CIOB Technical Information Service Paper No. 68, 1986.

HARPER, D., *Building. The Process and the Product*, Construction Press, 1978.

HERBERT, G., *Pioneers of Prefabrication*, The Johns Hopkins University Press, 1978.

HILLEBRANDT, P., *Economic Theory and the Construction Industry*, Macmillan, 1985.

INNOCENT, C.F., *The Development of English Building Construction*, David and Charles (Publishers) Limited, 1971.

JOHNSTON, J.E., *Site Control of Materials*, Butterworths, 1981.

JORDAN, P., *'QA: Simple Acronym Spells Big Changes', National Builder* (67) December/January 1986, pp. 358–360.

KING, J.E., *Upgrading Services in Domestic Properties*, Proceedings of CIBSE Seminar 'A Future for Existing Buildings – Domestic Refurbishment', 16th October 1986, pp. 14–19.

LEMESSANY, J. and CLAPP, M.A., *Resource Inputs to Construction: The Labour Requirements of Housing Building*, Building Research Establishment Current Paper 76/78.

MACPHERSON, I., 'Faster and Faster Track', *Building* (20) 15th May 1987, pp. 64–65.

MARTIN, B., *Joints in Buildings*, George Godwin Limited, 1977.

MOORE, C.E., *Concrete Form Construction*, Van Nostrand Reinhold Company, 1977.
NAGARAGAN, R., *Standards in Building*, Pitman Publishing, 1976.
NEDO, *Faster Building for Industry*, National Economic Development Council, 1984.
POWELL, M.J.V., *Quality Control in Speculative House Building*, CIOB Occasional Paper No. 11, 1978.
RIDOUT, G., 'Built Like a Shot. Paid on the Dot,' *Building* (252) January 30th 1987, pp. 34–36.
SCHUMACHER, E., *Small is Beautiful: A Study of Economics as if People Mattered*, Abacus, 1984.
SZOKOLAY, S.V., *Solar Energy and Building*, Architectural Press, 1978.
TOPLISS, C.E., *Demolition*, Construction Press, 1982.
VALE, B. AND R., *The Autonomous House*, Thames and Hudson, 1976.
WEEKS, I.G., *Site Foremanship*, Construction Press, 1978.
WELLS, D., 'Setting Sites on Quality', *Contracts Journal*, January 29th 1987, pp. 16–17.
WYATT, D.P., *Materials Management Part 1*, CIOB Occasional Paper No. 18, Chartered Institute of Building, 1978.
WYATT, D.P., *Materials Management Part 2*, CIOB Occasional Paper No. 23, Chartered Institute of Building, 1983.

Statutes

The Building Regulations, 1985, and Approved Documents.
The Construction (Lifting Operations) Regulations, 1961.
The Construction (Working Places) Regulations, 1966.
The Construction (Health and Welfare) Regulations, 1966.
The Construction (General Provisions) Regulations, 1961.
Health and Safety at Work etc. Act. 1974.

Index